An Everyday [illegible] of the Glob[illegible]

Most geographical studies of the 'Third World' – or the Global South – focus their attention on the challenge of promoting development and explaining why the Third World is also the Poor World. This text extracts the Global South from the shadow of development and examines people's lives and livelihoods in their own terms. It takes as its point of departure the need to reveal the myriad ways that people 'get by' in the day-to-day sense of the term and how modernisation is reworking the human landscape.

An Everyday Geography of the Global South focuses on local spaces, individual experiences, household strategies and the power and role of agency over structure in terms of explanation. Taking a broad perspective of liveliho[illegible] draws on more than 90 case studies from 36 countries across Asia, Africa and [illegible] people are engaging and living with modernity. This exten[illegible] households operate, to how and why people take on new [illegible] tion and mobility have become increasingly common [illegible] nd expectations are being reworked under the influence [illegible]

To date, there is no book which take[illegible] rstanding of the Global South. In focusing on the G[illegible] nning with the personal and the everyday, [illegible] illuminate and inform mainstream debates in [illegible] xperiences of 'ordinary' people, the book pro[illegible] o a range of geographical debates. For stude[illegible] of argument, its use of detailed case studies to [illegible] gument, and in providing a geography text whic[illegible] the Global South.

Jonathan Rigg is Professor of Geography at Durham University. His research interests include development in the South-East Asian region, rural and agrarian change, and political ecology.

An Everyday Geography of the Global South

Jonathan Rigg

LONDON AND NEW YORK

First published 2007
by Routledge
2 Park Square, Milton Park, Abingdon, Oxon OX14 4RN

Simultaneously published in the USA and Canada
by Routledge
270 Madison Ave, New York, NY 10016

Reprinted 2008

Routledge is an imprint of the Taylor & Francis Group, an informa business

Typeset in Baskerville by
Keystroke, 28 High Street, Tettenhall, Wolverhampton
Printed and bound in Great Britain by
MPG Books Ltd, Bodmin

British Library Cataloguing in Publication Data
A catalogue record for this book is available from the British Library

Library of Congress Cataloging in Publication Data
A catalog record for this book has been requested.

ISBN10: 0–415–37608–4 (hbk)
ISBN10: 0–415–37609–2 (pbk)
ISBN10: 0–203–96757–7 (ebk)

ISBN13: 978–0–415–37608–2 (hbk)
ISBN13: 978–0–415–37609–9 (pbk)
ISBN13: 978–0–203–96757–7 (ebk)

For Joshua, Ella, Cesca and Sam

Contents

Illustrations

Figures

Tables

Boxes

Preface

I decided to write this book for the simple and fairly standard reason that I felt that there was a gap in the literature and, therefore, in the market. More particularly, I felt that there was no university-level geography textbook that focused on the Global South – the 'developing world' – without also taking 'development' as its point of departure. As I explain in more detail in the opening chapter, it is often hard to think about or imagine the non-Western world without, at the same time, conjuring up visions of poverty, underdevelopment, inequality, and so forth. I wanted to extract the Global South from the tyranny of the development discourse and to examine the myriad geographies of the majority world unshackled from such associations. In this effort, I have to admit, I have been only partially successful. Try as I might, development seems to have found its way into the discussion, seeping in at every turn, through every fissure that words allow.

A second objective – and this was also driven by my sense that there was a gap in the literature – was to write a book that was about everyday lives. I wanted the focus to be on individuals, households and communities, rather than on governments, corporations and international organisations. The book therefore privileges the local, the everyday and the personal. Analysis has to start somewhere, from some vantage point, and in this instance I have intentionally chosen to begin at the roots of human lives and activities. In doing this, however, I am not suggesting that we can also end as well as begin with the personal and the everyday; lives and livelihoods are increasingly implicated in wider and 'higher' structures and processes. I return to this issue of linkage and interconnection in the final chapter but trust that the sense of everyday lives being, at the same time, global lives also comes through in the core discussion.

Inevitably, an author brings his or her own experiences and intellectual baggage – including knowledge – to bear in their work. My own specialist area of interest is the rural geographies of South-East Asia, and all my fieldwork has been undertaken in the countries of this region. I have no doubt that the experience of working in Asia has coloured my views and influenced my thinking. It could not do otherwise. However, the book draws on case studies and literature from across the Global South – not equally, to be sure, but pretty broadly. In total, more than 90 case studies from 36 countries are used to illustrate the discussion and substantiate the arguments.

A book of this type, which aims to straddle the world, is also at risk of simplifying the world. I have tried to avoid this by drawing few hard-and-fast conclusions, instead highlighting the reasons why contingency and indeterminacy so often rule the day. This should not be taken to mean, however, that we cannot 'learn lessons' from particular cases in specific contexts. It is possible to recognise diversity and difference while also searching out the patterns that make the world understandable. It is these patterns, the grammar that makes living decipherable, which I have been at pains to illuminate.

Jonathan Rigg
Durham

Acknowledgements

In writing this book I have had to venture outside my regional comfort zone of Asia and, as a result, have had to ask scholars with greater knowledge and experience than my own to cast a critical eye over my sometimes rather clunky distillations of their work. In this regard I am grateful to William G. Moseley (Mali), Rebecca Elmhirst (Indonesia), Kate Hampshire (Burkina Faso), Ann Le Mare (Bangladesh), Kate Gough and Paul Yankson (Ghana), Gina Porter (Ghana), Cheryl McEwan (South Africa) and Babette Resurreccion (Philippines). Less directly, there are many scholars, particularly across Asia, with whom I have been fortunate enough to work over the years and who – though they may not know it – are represented here in terms of the ideas and views which inform the discussion. Rather than listing them in turn, and running the risk of overlooking some people in what would end up being a long list, I have opted instead for a blanket statement of sincere thanks and appreciation.

Much academic work is collective work in the loose sense of the term, and this book is no exception. I have been extremely lucky to work with many fine scholars. These collaborations have ranged from joint authorship to co-editing, reviewing, teaching, research project management and paper presentations. But perhaps the most educational for me have been the opportunities I have had for joint fieldwork. Seeing other scholars in the field, observing how they engage with people, the questions they ask, the research methods they use, and the way at the end of the day they piece together disparate threads of evidence to build up an explanatory picture has been highly instructive. I sometimes think that the demands of fieldwork help to keep scholars human; in my case, it means that I can also count these scholars as friends.

At a more practical level, I would like to thank Catherine Alexander for her help in tracking down some more obscure journal papers on particular topics; Zoe Kruze, who acted as the editor on the book at Taylor & Francis through to July 2006; Jennifer Page, who took over from Zoe; and Andrew Mould, who commissioned the book in the first place. This is the third book of mine that has been through Andrew's careful hands and critical eye, and I am grateful for his continuing support, encouragement and confidence.

Most of the photographs in the book are my own, but Rebecca Elmhirst, Ann Le Mare and Kate Gough have generously permitted me to use some of their images. The figures and maps were expertly drawn by Chris Orton and Steven Allan in the Design and Imaging Unit of the Department of Geography at Durham and once again I have been lucky to benefit from their professional eyes and quiet efficiency. The manuscript was sent to three anonymous readers and their thoughtful comments were valuable in tying up some loose ends and in helping set out the broad structure for the final chapter.

Until now, my wife has sensibly refrained from reading my work, and my children have also shown no great inclination to do so. As each chapter of this book was completed, however, Janie gamely offered to cast an eye over the offering, perhaps guilty that it has taken her more than two decades and six books to venture past the cover and title pages of anything else I have written. Even so, and deep down, I don't think she harbours great hopes that this book will do much more than create a barely noticeable ripple of recognition. It would be nice to surprise her.

Acronyms and abbreviations

CBO	community-based organisation
CFG	Community Forest Group
CGIAR	Consultative Group on International Agricultural Research
DALY	disability-adjusted life year
DFID	Department for International Development (UK)
DRC	Democratic Republic of the Congo
EPI	Expanded Programme on Immunization
FAO	Food and Agricultural Organization (of the UN)
GDP	gross domestic product
GO	grassroots organisation
HDI	human development index
IFAD	International Fund for Agricultural Development
IIED	International Institute for Environment and Development
IRAP	Integrated Rural Accessibility Planning
IUCN	World Conservation Union
JFM	Joint Forest Management
LLDC	least developed country
NEM	New Economic Mechanism (Laos)
NGO	non-governmental organisation
NTAX	non-traditional agricultural export (crops)
OECD	Organization for Economic Cooperation and Development
PAR	participatory action research
PIP	Public Investment Plan
PO	people's organisation
PPA	participatory poverty assessment
PRA	participatory rapid appraisal
PRSP	Poverty Reduction Strategy Papers
QoL	quality of life
RRA	rapid rural appraisal
SIDA	Swedish International Development Agency
SL	sustainable livelihoods
SLA	sustainable livelihoods analysis
3-D jobs	dirty, degrading and dangerous jobs
TVEs	Township and Village Enterprises
UNDP	United Nations Development Programme
VO	voluntary organisations

Case studies

Chapter 1

Location	*Topic*	*Source*
Laos	Mrs Chandaeng and making a living	Rigg 2005
Vietnam	Environmental regulation	O'Rourke 2004
Indian Ocean – India, Indonesia	Tsunami and gender	Oxfam 2005
Ghana	Place and social capital	Porter and Lyon 2006
Pacific coast of Colombia	Sense of place among black communities	Oslender 2004

Chapter 2

Location	*Topic*	*Source*
Yunnan, China	Politics and bureaucracy	Pieke 2004
Brong Ahafo, Ghana	Women, forests and livelihoods	Leach et al. 1999
Qwaqwa, South Africa	Differentiation and diversification	Slater 2002
Burkina Faso	Participation and scale	Michener 1998
Mali	Cotton farming and inequality	Moseley 2005
Botswana and South Africa	Livelihoods and asset mobilisation	Twyman et al. 2004
Bangladesh	Well-being	Camfield et al. 2006

Chapter 3

Location	*Topic*	*Source*
Hanoi, Vietnam	Social capital	Turner and Phuong An Nguyen 2005
Ghana and Senegal	Social capital	Lyons and Snoxell 2005
Guamote, highland Ecuador	Social capital	Bebbington and Perreault 1999
South Sumatra	Livelihoods and the life course	Leinbach and Del Casino 1998; Leinbach and Watkins 1998
Caribbean and Indonesia	Life course and work	Grijns and van Velzen 1993; Momsen 1993
Thailand	Cultures of modernity	Mills 1997, 1999
Zambia	Cultures of modernity	Ferguson 1999; Cliggett 2003
Nepal	Cultures of modernity	Pigg 1992, 1995, 1996
Kerala, India	Consumerism and consumption practices	Osella and Osella 1999

Chapter 4

Location	*Topic*	*Source*
Central America	Commercial activities in pre-modern period	McSweeney 2004
Borneo	Commercial activities in pre-modern period	Cleary 1996, 1997
Ghana	Livelihood transitions in rural areas	Kunfaa 1999; Kunfaa and Dogbe 2000; Whitehead 2002
Ghana	Land markets and livelihoods in peri-urban Accra	Gough and Yankson 2000
Dhaka, Bangladesh	Livelihood transitions in Dhaka (rickshaw pullers)	Narayan et al. 2000; Begum and Sen 2004, 2005; Conticini 2005
Mongolia	Livelihood transitions and reform	NSOM 2001; Mearns 2004
Kinshasa, Democratic Republic of the Congo	Livelihood transitions and chronic insecurity	Iyenda 2005
Bangladesh	Livelihood transitions and flooding	Nabi et al. 1999, 2000; Ninno et al. 2002
Khao Lak, southern Thailand	Livelihood transitions and the 2004 tsunami	Rigg et al. 2006
Sri Lanka	Livelihoods and civil war	Koff 2004

Chapter 5

Location	*Topic*	*Source*
Tianjin, China	Individual modernities	Inkeles et al. 1997
Mexico	Global agro-food systems	Sanderson 1986; Friedmann 1993; Raynolds 1993; Kearney 1996; Barkin 1990, 2002
Guatemalan highlands (municipalities of Tecpán and Santa Apolonia)	Non-traditional agricultural export crop production	Hamilton and Fischer 2003
Kenya	Cut flower production	Hale and Opondo 2005
South Africa	Working in the fruit export industry	Kritzinger et al. 2004
Bobo-Dioulasso, Burkina Faso	Non-traditional agricultural crop production	Freidberg 2003
Ayutthaya, Thailand	Consumer lifestyles	Rigg (current fieldwork)
Sinos Valley, Brazil	Shoe factory work	Schmitz 1995
Lampung, South Sumatra and Tangerang, Jakarta, Indonesia	Female migration and factory work	Elmhirst 1995a, 2002, 2004
Banjaran, Indonesia	Female factory workers	Hancock 2000, 2001
Perak, Malaysia	Malaysian *kampung* as socially urban spaces	Thompson 2002, 2003, 2004
Pearl River Delta, Guangdong province, China	Confucianist traditions and modern factories	Wright 2003

Chapter 6

Location	*Topic*	*Source*
Puruliya District, Chottanagpur Plateau, West Bengal, India	Migration, mobility and historical change	Rogaly and Coppard 2003
Rural Nepal and India	Migration, livelihoods and policy-making	Deshingkar and Start 2003; Gill 2003
Madhya Pradesh and Andhra Pradesh, India	Migration, livelihoods and diversity of outcomes	Deshingkar and Start 2003
Slums of Delhi, India	Migration, livelihoods and diversity of outcomes	Mukherjee 2004
Papua New Guinea	Mobility as intrinsic and inherent	Curry and Koczberski 1998

Melanesia	Mobility as intrinsic and inherent	Chapman 1995
Papua New Guinea	Migration and identity	Curry and Koczberski 1999; Koczberski and Curry 2004
Wuwei County, Anhui Province, China	Female migration and rural–urban relations	Yan Hairong 2003
Northern Burkina Faso	Fulani and *exode*	Hampshire 2002
Bihar, India	Migration and the 'left behind'	de Haan 2002
Bengal, India	Migration and the historical shaping of gender relations	Sen 1999
Java, Indonesia	Migration and the co-constitution of rural areas and mobility	Breman and Wiradi 2002
Java, Indonesia	Migration and return	Koning 2005
Zambia	Counter-urbanisation and reverse migration	Potts 1995, 2000, 2005
Ifugao, Philippines	Migration and 'remittance landscapes'	McKay 2003
Thailand, Indonesia	Rural areas and source communities as 'safety nets'	Silvey 2001; Rigg 2002, 2003
Western Isles of Scotland	Rural out-migration and community sustainability	Stockdale 2004

Chapter 7

Location	*Topic*	*Source*
Laos	Land allocation and land settlement	Vandergeest 2003; Rigg 2005
Kalimantan and Sulawesi, Indonesia	State orchestrated resettlement programmes	Li 1999, 2005
Java and Sumatra, Indonesia	Decentralisation	Bebbington et al. 2004
Cusco, Peru	Role of NGOs	Bebbington 2005
India	Participation and participatory exclusions in the Joint Forestry Programme	Corbridge and Jewitt 1997; Agarwal 2001
Thailand	Participatory exclusions in irrigation schemes	Resurreccion et al. 2004
West Bengal, India	State–society relations and local government reforms	Williams 1997

Western Cape and KwaZulu Natal, South Africa	Citizenship and everyday democracy	McEwan 2003, 2005
Ethiopia and Mozambique	Natural resource management	Black and Watson 2006
Andhra Pradesh and Madhya Pradesh, India	Decentralisation	Johnson et al. 2005
Namibia	Decentralisation of rural water supply	Thomas and Twyman 2005
Vietnam	State–society relations and everyday politics	Kerkvliet 1995b, 2005
Guangzhou, China	State controls and the *hukou* system	Fan 2001, 2002
Karachi, Pakistan	Informal housing and state–society relations	Hasan 2002
Java, Indonesia and Saudi Arabia	Female domestic transnational workers	Silvey 2004
Fiji	Colonial systems of government and control	Thomas 1994

Chapter 8

Location	*Topic*	*Source*
Ceylon (Sri Lanka)	Resistance in nineteenth century coffee plantations	Duncan 2002
Malaysia	Resistance in electronics factories	Ong 1987
Thailand	Resistance, religion and tree ordination	Isager and Ivarsson 2002
Vietnam	Resistance in the Vietnamese countryside	Kerkvliet 2005
China	The state and resistance in China	Cai 2004
Mexico	Resistance as anti-globalisation	Barkin 2002
Colombia	Place-based resistance among Black communities	Escobar 2001
Pakistan	Place-based resistance and global resistance networks	Butz 2002
Colombia and Ecuador	New social movements and anti-globalisation	Bebbington 2004a; Escobar 2004
Colombia	The creation of 'Indianness'	Jackson 1995
Sulawesi, Indonesia	Resistance movements and hydropower	Li 2000, 2001

Pakistan	Resistance in the third space of trekking trails and portering	Butz 2002
Philippines	Gender, identity, forest management and livelihoods	Resurreccion 2006

Chapter 9

Location	*Topic*	*Source*
Indonesia	Community planning	Beard 1999, 2002, 2003
Syria	Feminisation of agricultural labour	Abdelali-Martini et al. 2003
Laos	Mrs Chandaeng	Rigg 2005

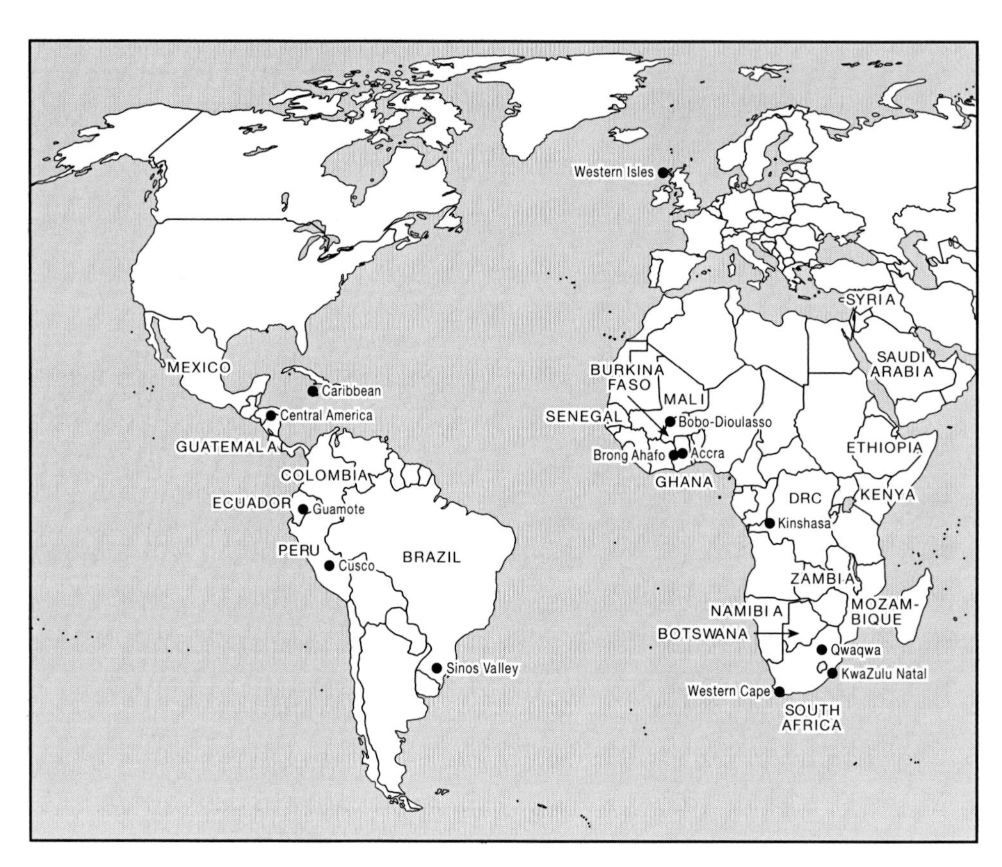

Map 1: Location of case studies referred to in the text

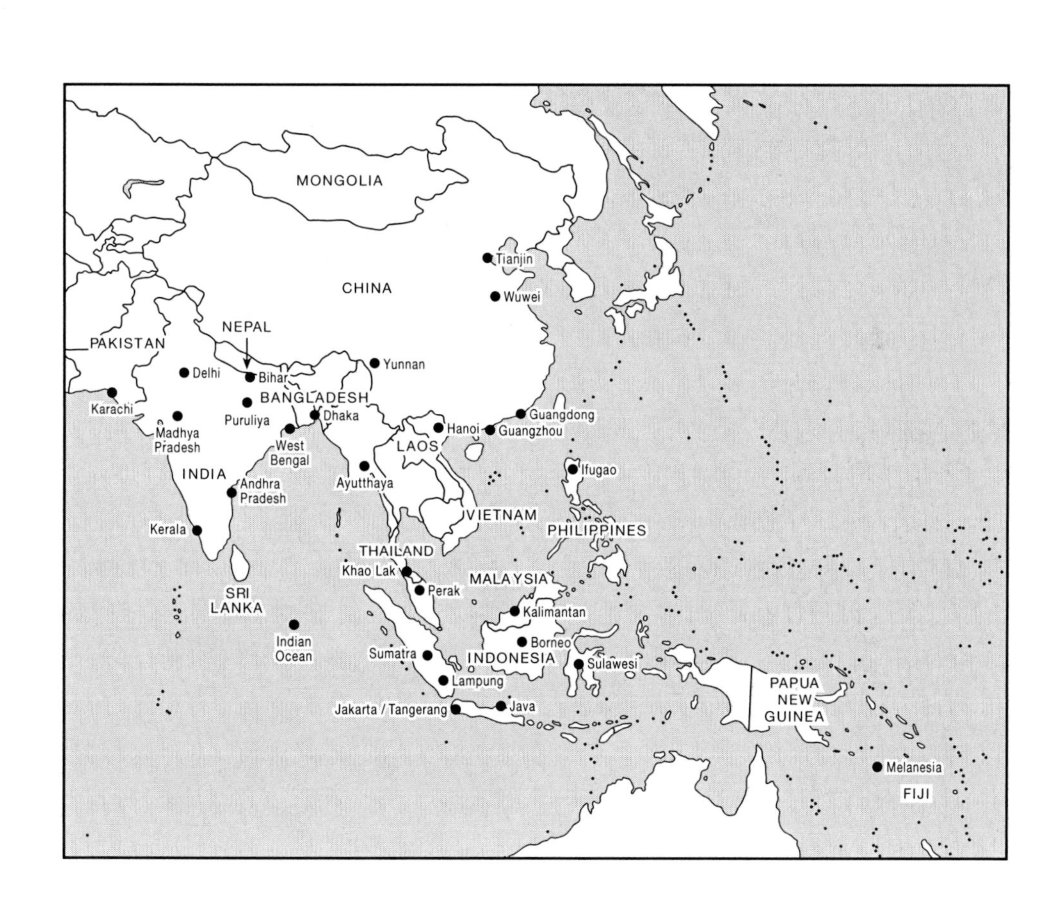
MONGOLIA
CHINA
Tianjin
Wuwei
NEPAL
PAKISTAN
Yunnan
Delhi
Bihar
BANGLADESH
Karachi
Puruliya
Dhaka
Guangdong
Madhya Pradesh
Hanoi
Guangzhou
West Bengal
LAOS
INDIA
Ayutthaya
Ifugao
Andhra Pradesh
VIETNAM
PHILIPPINES
Kerala
THAILAND
Khao Lak
MALAYSIA
SRI LANKA
Perak
Kalimantan
Indian Ocean
Borneo
Sumatra
INDONESIA
Sulawesi
Lampung
PAPUA NEW GUINEA
Jakarta / Tangerang
Java
Melanesia
FIJI

1 What's with the everyday?

The everyday, globalisation and the Global South

Mrs Chandaeng: an everyday geography

We met 50-year-old Mrs Chandaeng on a cool December day in 2001 outside her house in the village of Ban Sawai in Laos, one of the world's 49 'least developed' countries (Illustration 1.1). She had been born and raised in the war-torn province of Xieng Khouang, where she met and married her husband, Udom. They settled in his home village and had six children. In 1988, however, when their youngest daughter was just two, Udom died suddenly and after a dispute with her brother-in-law, Mrs Chandaeng moved to Ban Sawai, settling there with her young family in 1991. As a newcomer, Mrs Chandaeng was unable to secure any land beyond her house plot and, in the context of a village economy based on farming, she struggled to feed and raise her six children. Yet when we interviewed Mrs Chandaeng ten years after she had first settled in Ban Sawai she was in the process of building a new and

Illustration 1.1 Village scene, northern Laos

impressive house. Her ability not only to survive but, ultimately, to prosper as a landless, widowed mother of six was surprising given what we knew of structural patterns of poverty and prosperity in rural Laos. Landless, female-headed households, and particularly those with young families, are usually among the poorest in an already very poor country.

It quickly became clear why Mrs Chandaeng had managed to buck the trend: four of her children were working in neighbouring Thailand, remitting between them around US$25–50 a month. Her son, Kai, was working on a shrimp farm while her three daughters, Wan, Lot and Daeng were employed as housekeepers in Bangkok. She may have explained her children's sojourns in Thailand in terms of 'when you are poor, you have to go', but the outcome was a degree of economic prosperity, at least in village terms.

The experience of Mrs Chandaeng represents in microcosm many of the core issues which this book attempts to confront and illuminate. First of all, to go beyond structures to understand the personal geographies which make ordinary people and their lives extraordinary. Second, to appreciate that beneath the summary data – the averages, means and aggregates – is a degree of difference and variability that challenges whether such summaries can be regarded as representative of the collective experience, and vice versa. And third, to see livelihoods as becoming increasingly implicated in geographies of globalisation but in ways in which people like Mrs Chandaeng become more than mere objects of – and for – development, but subjects with their own volition.

An everyday geography of the Global South

A criticism that has been levelled at geography over recent years is that it would seem to have forgotten the importance of taking a truly global perspective and replaced an interest in place (and places), with a seemingly ever more abstract concern for space. This is a reactionary viewpoint to hold. Place-based geographies are associated with regional geography and, as we all know, regional geography is just so yesterday. Worse still, regional geography smacks of colonial geographies, and colonial geographies of domination, control, and worse. This book is not, I hope, just a throwback to an earlier geography but an attempt to present a different and, to some degree, an alternative geography. It is different and alternative in three main senses.

First of all, the book is explicitly about the 'Global South' (see Box 1.1). An alien leafing through recent issues of mainstream English language human geography journals might think that the countries of the non-Western world were a mere adjunct, a small and rather dry annex, to the West. For example, of 362 papers published between 2003 and 2006 in three of the most influential geography journals – *Progress in Human Geography*, *Transactions of the Institute of British Geographers* and the *Annals of the Association of American Geographers* – fewer than one in eight had a primary and explicit concern with countries, conditions or processes in the Global South.[1] Part of the reason for this may be that relatively few scholars based outside the rich Anglophone world publish in such journals. Gutiérrez and López-Nieva in a review of the content of 19 English language human geography journals note that 74 per cent of articles were written by scholars based in the United Kingdom or the United States (Gutiérrez and López-Nieva 2001). This dominance of Western/Northern scholarship, and the overriding focus on the geographies of the West/North, is a source of concern in the obvious sense that the Global South is important and, furthermore, is becoming more so as each year passes. In 2005 the output of the Global South exceeded half of total world gross domestic product (GDP); by 2025 this figure will likely be more than 60 per cent (The Economist 2006b: 3–4). For a whole range of political, economic and environmental reasons

BOX 1.1 Defining terms

Term	*Usage/definition and appropriateness/weaknesses*
Core/ periphery	The terms core and periphery (and semi-periphery) have a long history, coming into widespread use in academic circles following the publication of Immanuel Wallerstein's (1974, 1980, 1989) three volume work on the modern world system. Core and periphery are also used in terms of the loci of academic and political power.
Developed/ less developed	These terms refer to the level of economic development of countries and to the rich/poor world binary. They highlight continuing global inequalities but categorise countries as either 'developed' or 'less developed' when these two categories are internally highly differentiated. They also gloss over the degree of mobility at the margins where some fast-growing economies have made the transition from de facto less developed countries to developed countries, sometimes in less than a generation.
First World/ Third World	These two terms – and also the linked 'Second World' and 'Fourth World' terms – are a legacy of the Cold War and refer to a geopolitical divide between the capitalist/liberal democratic First World and a Communist Second World. The Third World was the residual but became quickly redefined as the poor world. The term Fourth World emerged rather later and variously refers to the tribal peoples of the world, (stateless) refugees or the world's least developed countries (or LLDCs). Most scholars avoid using the terms because they are historically obsolete and because of the perceived pejorative connotations associated with the terms 'First' ('best') and 'Third' ('worst') worlds. (See Berger 2004 for a discussion of the fate of Third Worldism.)
Global North/ Global South	This is a derivation of the North–South divide noted below in this box. Some scholars prefer to add 'Global' to make it clear that this is not a strict geographical categorisation of the world but one based on economic inequalities which happens to have some cartographic continuity. In addition it emphasises that both North and South are, together, drawn into global processes.
Majority world/ minority world	This turns the usual ordering of the binary (N–S, Rich–Poor, First–Third) on its head to make it clear that the South/Poor/ Third World is the majority world supporting some 80 per cent of the globe's population and 136 of the 192 recognised states.
North–South	The North–South distinction is associated with the Brandt report of 1980 (*North–South: a programme for survival*) which argued that 'in general terms, and although neither is a uniform or permanent grouping, "North" and "South" are broadly synonymous with "rich" and "poor", "developed" and "developing"' (Brandt 1980:

continued

North–South continued	31). The obvious deviations from the geographical categorisation of North/rich and South/poor are Australia and New Zealand. In addition, over the years since the report was published the internal coherence of a 'South' has become even more problematic (see Slater 1997).
Western world/ non-Western world	This Western/non-Western duality refers to a perceived cultural divide between, essentially, 'the West' and 'the Rest'. This is not synonymous with North/South or rich/poor for the reason that some countries of the non-Western world, notably Japan, are Northern and rich. The term Eurocentric or Eurocentrism is used to highlight the dominance of Western viewpoints in many areas of life and thought.

we need to know more about each other, and the North–South balance of academic knowledge and (apparently) interest is out of kilter. But it is also of concern in another sense: the papers betray a channelling and domination of Northern knowledge to, and over, the South. Conceptual and theoretical approaches and frameworks that have their roots in the North are used to frame and explain the South. Rarely does the flow of knowledge run counter to this stream and even more rarely is it seriously considered that the South might have something to teach the North. The assumption is that Northern geographies are relevant and appropriate for understanding the South:

> Seldom do those [geographers] who work in the core [countries of the North] recognise the particularities of their own geographies; that their theories do not travel, that their global geographies are partial and that developments in the core are interdependent with the periphery.
>
> (Bradshaw in Olds 2001: 133; see also Yeung and Lim 2003)

This issue has been most vigorously pursued in Dipesh Chakrabarty's (2000) book *Provincialising Europe*. Charkrabarty argues that Europe, for historians, acts as a silent referent: all Third World historians are required to touch their forelock and comment on European history and scholarship, but not vice versa. His book is not, however, a call to reject European scholarship as an act of 'postcolonial revenge'. Western scholarship is both 'indispensable and inadequate' to the task of explaining the non-Western world (Chakrabarty 2000: 16).

In their review of globalisation and the geography of cities in the less developed world, Grant and Nijman (2002) conclude that their 'study shows that the economic geography of present-day cities in the less-developed world is fundamentally different from that of "Third World cities" *and* from globalizing cities in the West' (2002: 328, emphasis in original; see also Robinson 2002). In other words, not only is the flow of knowledge predominantly one-way (undesirable in itself), but also this can lead one to the false assumption that in this era of globalisation all processes, dynamics and their outcomes are, essentially, the same. Using Mumbai (India) and Accra (Ghana) as their case studies, Grant and Nijman show that place (i.e. the idiosyncrasies of location) and history are critical in understanding how globalisation processes operate (Illustration 1.2). Robinson (2002) makes much the same accusation in her article on the world cities literature in which she writes that 'theoretical reflections should at least be extremely clear about their limited purchase and, even better,

Illustration 1.3 Recycling on the outskirts of Hanoi, Vietnam. Recyclers are often drawn from particular villages around Hanoi and specialise in recycling certain products

Starting with, or privileging, the local and the everyday is important not only because it highlights the explanatory distance separating macroscopic and microscopic interpretations, but also because it forces a consideration of human agency. And when human agency enters the explanatory fabric, contingency, serendipity and personality do so too. Global perspectives have been criticised for their 'abrogation of agency' (Ley 2004: 154) with a resultant tyranny of process that leaves little space for difference. As is explored in greater detail below, the devil – and the fascination – really is in the detail of globalisation.

There is also a third feature of the book which requires noting and some elaboration. The Global South is also known, variously, as the Third World, the poor world, the less developed world, the non-Western world, and the developing world (Box 1.1). The term Global South is just the latest in a cavalcade of terms that seek neatly to package and pigeonhole the world that is beyond our own.[3] As these alternative terms make clear, the Global South also invariably becomes recast or repackaged in poverty/development terms. It is hard even to begin the process of thinking about the Global South without also irrevocably linking it with the challenge of development and the stain of poverty. This book, however, while it is about the Global South is not, in the first instance, about development, nor about poverty – although the poor, and their struggle to survive and to prosper, are inevitably among the cast of characters. This is partly for the reason that there are many excellent development texts. However in this era of globalisation, it has been argued that all countries are developing and there is a convincing case that the essential *qualitative* differences between, say, Belgium and Burkina Faso, or Canada and Colombia are not to be found rooted in each countries' different level of development. Indeed, the focus on differences in levels of development or modernisation have, it could be argued, distracted attention from other, even more profound, manifestations of difference. To be sure, the fact that the incidence of poverty, using the World Bank's US$2 a day cut off, is zero in Belgium but 81 per cent in Burkina Faso contains a pressing message about global inequality. But if one were to set about writing a book about Belgium or Canada it is unlikely that it would be framed in terms of development. It also raises the question of what happens when formerly poor countries bridge the economic gap with the rich world. Self-evidently, the high performing economies of East Asia – Japan, South Korea, Taiwan, Singapore and Hong Kong – have not analytically merged with the Western world simply because their per capita incomes are similar. Perhaps it is for this reason that the United Nations continues, rather unsatisfactorily, to classify these latter four economies as 'emerging'.

This book, then, also rests on the premise that there are important things to say and understand about the Global South that are independent of the challenge of propelling development and reducing poverty. In writing this I am also implying that the ever-present focus on poverty/development takes something away, whether by distraction or omission, from our view of the Global South. Pictures of huddled masses, peasants working their land, or young men and women in torn shorts and T-shirts picking over steaming garbage mountains are certainly evocative and a powerful cry to confront global poverty. They also, though, contribute to the victim status of the Global South and deny these people the dignity that comes from at least some degree of agency and independence of action. Ordinary people in the Global South are, like everyone else, extra-ordinary and to appreciate this requires that they become more than objects to be 'developed'.

Making space for the 'everyday' in an era of globalisation

> It is geography, perhaps, that is confronted by the potentially most destabilizing implications [of globalisation], for according to some commentators globalization is expunging local difference and hence the relevance of space and place.
>
> (Martin 2004: 148)

The fears and expectation expressed in the quote above are well known: globalisation is leading inexorably to a borderless world where cultural homogenisation, media imperialism, transnational domination and economic integration are propelled and controlled by the expanding tendrils of information and communications networks, a global financial architecture, and an increasingly powerful phalanx of multilateral institutions. Deterritorialisation and the so-styled 'death' of geography is the outcome.

Yet, just at the time when globalisation has become the defining process of the age and some scholars – indeed some geographers – are sounding the death knell of geography, there has emerged a vibrant concern for the minutiae and distinctiveness of the 'everyday' and, by association, the local. The 'everyday', it would seem, is everywhere. Scholars write of 'everyday political practices' (Flint 2002: 391), 'everyday resistance' (Scott 1985), 'everyday urban travel' (Røe 2000), 'everyday food' (Freidberg 2003), 'everyday politics' (Kerkvliet 2005) and, more widely, of 'everyday life' (Appadurai 1999; Simard and De Koninck 2001; Duncan 2002), 'everyday lived realities' (Kabeer 1997) and 'everyday geographies' (Oslender 2004). There is, furthermore, an increasing recognition that the everyday is often the critical component in building an understanding of the processes underway. How can we reconcile the surface contradiction between the emergence of a world worn flat by the indefatigable forces of globalisation, and a world where localities and localism are gaining in significance and where difference and complexity are becoming ever more pronounced and powerful?

This contradiction between an intensifying process of globalisation and a growing concern with the power of the everyday is partially explained by the recognition that globalisation itself is a deeply contradictory process. The fears and expectations of the early hyper- or ultra-globalists have been tempered by the realisation that globalisation has not erased the local and the everyday but, often, re-energised it. Globalisation operates at all levels and scales simultaneously, and the relationships can operate both ways (Yeung 1998; Kelly 2000). Globalisation can strengthen local regulation, bolster and empower local groups, strengthen and revivify local cultures, while localities can both shape and respond to global processes. These alternative perspectives challenge the all too common tendency towards the pigeon-holing of globalisation as a hegemonic and totalising force issuing from the core and colonising and dominating the periphery. Such a view not only situates globalisation as a process with its origins in the Global North, but also identifies the trajectory of effect as one that leads from the Global North to the Global South. As Rapley says of development theory and practice, but which resonates with debates over globalisation:

> much development thought was not imposed on the developing world by the developed world, but rather emerged from [the developing world]. . . . Equally, much of the resistance to development comes not from 'traditional areas' [i.e. the Global South], but from urban activists in the First World.
>
> (Rapley 2004: 351)

The debate and the political and social tensions that have arisen over China's growing role in the world economy is but the most visible aspect of this 'retuning' of globalisation.

There are a variety of ways in which scholars have sought to challenge the 'death of geography' thesis. To set these out in summary:

- Globalisation operates at all scales. It is not that the global is in the process of erasing the local, but that globalisation processes can be seen operating at the local scale (and at all other scales up to the global). Globalisation does not, therefore, lead to eradication, but transformation.
- Globalisation, scale and the straitjacket of the local/global binary can be partially reconciled if 'rather than viewing the local and global, or place and space, as distinct scales, they [are] instead seen as "nested"' (Kelly 1999c: 153). (This is discussed in more detail in the next section.)
- Globalisation – like capitalism – is uneven in its effects and therefore, by definition, there will be a geography to globalisation. While global capital may be increasingly mobile, it does not go everywhere. Some places are 'stickier' than others, creating disparity and unevenness (see Glassman 2002). Geographers therefore have the task of describing and

explaining the patterns that emerge and these will have a geography that goes beyond, and before, the global.

- There is a meta-narrative of globalisation – as there is of modernisation (discussed below) – but this does not mean that either globalisation or modernisation are singular processes. Globalisation leads to multiple modernities and social scientists need to identify, interpret and attempt to understand these multiple, and often contradictory, outcomes (see Englund and Leach 2000).
- Globalisation has, itself, played a part in creating the political and institutional space for local, grassroots initiatives from devolved systems of government to locally rooted non-governmental and people's organisations. This has led some people to talk of a parallel process of 'relocalisation'. The local, in this way, becomes a project rather than a fact.
- More widely, there is a strong case for arguing that globalisation is always implicated and embedded in the cultural and historical context of a place, a community, a household and an individual. It becomes inevitably particularised. Thus, for example, women's incorporation into the global labour force in the context of factory work is moulded both by the forces of economic restructuring driven by the New International Division of Labour and by place-based geographies which are culturally and historically as well as geographically contingent (see, for an example, Silvey 2003).
- Even cosmopolitans and cosmopolitanism are situated (see Ley 2004). This applies not only to patrician (or privileged) cosmopolitans – such as the expatriates and multinational white collar employees who are normally associated with the term – but also to the plebeian cosmopolitans who may service those higher up the employment and income ladders. These include, for example, Sri Lankan and Filipino domestic workers in Saudi Arabia and Singapore, and local rural migrants working in the homes of public sector employees in India.

This gradual reframing of globalisation is not to deny the power of the forces that the process has unleashed. The market *is* infiltrating even the most remote of regions and hitherto the most self-reliant of communities. Consumerism, defined by the consumption of certain global brands, *is* in the process of becoming the watchword of the many, rather than the luxury of the few (Illustration 1.4). Technology *is* levelling the playing field. The role of certain pivotal multilateral organisations *is* inexorably growing. And governments *are* finding their room for manoeuvre in an increasingly interlinked and interdependent world hampered and constrained. The thesis that Thomas Friedman expounds in his widely read book *The world is flat: a brief history of the globalized world in the twenty-first century* (2005) is a convincing one, not least because it is, apparently, all around us and there to be seen and checked against our experience:

> I just wanted to understand why the Indians I met were taking our [American] work, why they had become such an important pool for the outsourcing of service and information technology work from America and other industrialized countries. [Christopher] Columbus had more than one hundred men on his three ships. . . . When I set sail, so to speak, I too assumed the world was round, but what I encountered in the real India profoundly shook my faith in that notion. Columbus accidentally ran into America but thought he had discovered part of India. I actually found India and thought many of the people I met there were Americans. . . . Columbus reported to his king and queen that the world was round. . . . I returned home and shared my discovery only with my wife, and only in a whisper. 'Honey,' I confided, 'I think the world is flat.'
>
> (Friedman 2005: 4–5)

Illustration 1.4 Middle class consumers outside Starbucks in central Bangkok, Thailand

While I do not work with the computer analysts and call centre operators that Friedman interviews, I too am struck, year after year, by the way in which these various processes of globalisation insinuate their way into local livelihoods and everyday lives. I can point, for example, to cassava farmers in the poor north-eastern region of Thailand who have regaled me with their views on the trade policies (and injustices) of the European Union; young women in villages in northern Thailand telling me of their desire to work in 'clean and modern' factories so that they can buy a motorbike on hire-purchase and become 'up-to-date' or *than samai*; parents in villages in rural Laos who have made clear their driving desire to educate their children so that – above all else – they can have a better life than their own; and to children from peasant families in Sanpathong, also in northern Thailand, streaming into internet shops in the district town after school so that they can engage with a world beyond their own. Experiences such as these provide a drum beat of proof to support that nagging sense that the world, hitherto so richly variegated and uneven, is being worn flat by globalisation. However accounts like Friedman's, I would argue, have the effect of playing down and often ignoring the uneven geographies of globalisation in their desire to tell a good story. Culture becomes in thrall to economics, and the social in thrall to technology.

This book, however, tries to take our understanding of globalisation 'down' to another level of detail. It is here, at the level of the everyday, where the 'worn flat' thesis begins to fray. What does it mean to be up-to-date in northern Thailand, or in Brazil, or in Ghana? They may consume, like the rest of us, but *how* do they consume? How do farmers in the Global South resist – or embrace – the trade policies of the Global North? What are the values that inform parental decision-making? At the meso-level it may appear that the countries and populations of the Global South are playing the same game of catch-up with the same rules and goals. But when we shift to the micro-level it becomes clear that there are variants of games, each with their own rules and their own nuanced goals. Some scholars see this as a disjuncture, implying that it is a puzzle and pregnant with contradiction (see Schirato and Webb 2003: 15–20). It is also one of the reasons why some people studiously avoid using the word 'globalisation', because it would seem to be a trap and a dead-end. But, as Schirato and Webb (2003: 19) say, 'if a term is being used so often, in some many different theatres, and with such profound effects, it is worth paying attention to it'. Rather than focusing on the disjunctures and the contradictions of globalisation, this book aims to elucidate and illuminate the micro, the local and the everyday within the wider ambit of globalisation.

Space, place and scale: the practice of theory

Central to geography from the beginning has been a concern for space, place and scale. A great deal of theoretical ink has been spilt – and continues to be spilt – by geographers seeking to unravel and interrogate these three critical terms. It will already have become clear, just a handful of pages into the first chapter, that the concerns of this book are situated at the local level and couched in terms of the spaces of the everyday. For some geographers this is deeply problematic from the very start because such an approach harbours an inbuilt assumption that there is a 'local' scale, distinct and separate from the 'non-local', and that there are everyday spaces of activity which have coherence and meaning in, and of, themselves. In this section I will attempt to explain – and justify – the approach that is taken.

First, though, an admission: theory does not come particularly easily to me; I do not revel in theory for its own sake, but rather see its utility in helping to structure and explain the patterns of the world. No doubt this is because much of my academic work begins with fieldwork: I engage with and seek to understand the hopes and desires, the challenges and problems, and the opportunities and tensions that farmers, factory workers, fisherfolk, craftsmen and women, students, and elderly people grapple with in different contexts and countries across Asia.[4] This is not a badge of honour or a claim for privilege. It is a simple statement of fact. The result, though, is that when I engage with theoretical debates over scale, space and place I do so with the furniture of field experience clogging the room.

In his work on the dialectics of space, Henri Lefebvre (1991) tried to overcome the space/place dichotomy or binary by reconciling both spatial scales, and physical and social space. In effect, his work brings together the local scale and everyday practices (social space) with the global scale and physical space. Global capitalism is not abstracted from the everyday, but grounds itself in specific places, in certain ways, and with particular social consequences (see Merrifield 1993). Space becomes both network *and* process, both place *and* flow.

> It follows . . . that place is not merely abstract space: it is the terrain where basic social practices – consumption, enjoyment, tradition, self-identification, solidarity, social support and social reproduction, etc – are lived out. As a moment of capitalist space, place is where everyday life is *situated*. And as such, place can be taken as *practiced space*.
>
> (Merrifield 1993: 522, emphases in original)

In his attempt to reconcile the space/place dichotomy, Lefebvre (1991) considered there to be three 'fields' of space, namely representations of space; representational space; and spatial practices. Notably absent is material space – space that can be measured in a mathematical sense. For Lefebvre, space is a product of social and political actions. Such actions do not populate physical space; they create their own space. But this space is not abstract; it is mental *and* material, and we can truly understand it only when the mental and material are combined (see Elden 2004). Finally, Lefebvre (1991) emphasises that it is important to appreciate that all spaces have a history and that they are continuously reconstituted as social and political actions evolve.

Table 1.2 outlines the essential distinctions between these three fields and provides examples of how such fields can be seen operating in practice (although in theory, all operate simultaneously). Merrifield (1993) sums up Lefebvre's approach to place by writing that it is 'the "moment" when the conceived, the perceived and the lived attain a certain "structured coherence"' (1993: 525; see also Merrifield 2000).

For Lefebvre (1991), the local or the micro-scale remains important as the site in which everyday practices are enacted. However, his work cautions against taking this to mean that the local is also the site in which relevant forces and influences will be deployed. The local is intimately networked into other scales.

Agnew and Duncan (1989: 2) provide – rather more simply – a threefold segmentation of place into location, locale and sense of place drawing on approaches to place that are characteristic of different disciplinary persuasions. *Location* is the classical geography of an area describing its physical components and the historical, economic and social processes that have made it. They link this with the work of economists and economic geographers. *Locale* is the rather more malleable and shifting context within which everyday living occurs and picks out those elements of location which are relevant and important to living. Locale, in effect, becomes a subset of location and is characteristic of the work of micro-sociologists and humanistic geographers. *Sense of place* refers to the ways in which the human imagination bestows on locations particular qualities, meanings and significance. This approach to place tends to be associated with cultural geographers and anthropologists. While each approach can be linked with different disciplines and subdisciplines, Agnew and Duncan (1989) argue

Table 1.2 A Lefebvrian practice of space

Lefebvre's spaces	*Types of space*	*Examples*
Representations of space	The space of planners, scientists and technocrats; 'expert', 'breaucractic' or 'elite' space. While this space can often be measured and bounded it is, nonetheless, an abstract space because it is linked to a set of (dominant) knowledges and skills. It is hegemonic space, ordered to fulfil the objectives of society.	State-orchestrated policies of territorialisation (see page 148); decentralisation policies that direct resources or control to lower levels of administration (see page 156); *hukou* policies in China which classify people as rural or urban residents (see page 123).
Representational space (or social space)	Lived space; the space of everyday experience where the spatial practices of everyday life and the routinised social relations of production and reproduction occur: factory, bedroom, theatre, street. Unlike the space of planners and technocrats (representations of space, as above) this is passive space which planners attempt to dominate and control: 'Lived space is the experiential realm that conceived and ordered space will try to intervene in, rationalize, and ultimately usurp' (Merrifield 2000: 174).	Community-level environmental regulation and management in Vietnam (see page 8); 'empowerment' in patriarchal households (see page 26); spaces of resistance in factories (see page 179) and plantations (see page 169).
Spatial practices	The activities and actions that structure everyday life consisting of networks, flows, patterns and routes and encompassing both production and reproduction.	Social action in China (see page 27); migration flows segmented by gender, generation or caste/ethnicity (see page 125); the participation of individuals in community activities and projects (see page 152); the politics of place (see page 178); and see Table 8.4.

that they should not be pursued separately, but regarded as complementary. Oslender (2004) has taken this threefold perspective on place and used it as a means of structuring his work on black communities living along Colombia's Pacific coast (Table 1.3). What is evident from Table 1.3 is that not only does each approach take a different viewpoint, but also it adopts a different way of looking.

What is meant by the 'everyday'?

In thinking about the role and place of the everyday, a number of other terms are shepherded into view: the commonplace and the ordinary, as well as (though less satisfactorily) the banal and the prosaic. Here the starting point is to focus (but not necessarily at the same time) on

- ordinary people
- everyday actions
- commonplace events.

Table 1.3 The places of the everyday: black communities on Colombia's Pacific coast

Agnew's 'places'	*Application to Colombia's Pacific coast black communities*
Location	10 million hectares; 1,300 km of coastline; high levels of precipitation; high levels of biodiversity; extensive network of rivers; prone to flooding; population of 1.3 million, 93% African Colombians; settlement concentrated along rivers; subject to boom-and-bust economic cycles . . .
Locale	Rivers provide the principal focus for living, livelihoods and life. Houses are constructed along rivers, transport is by river, and livelihoods are based on fishing and the collection of shellfish. Social interaction occurs in, on and along rivers, and the tide determines the rhythms of life. This is where people wash, play and gossip. Rivers also mediate interactions between communities along the length of the waterways.
Sense of place	The river becomes, because of its centrality to existence, a 'central point of reference in identity formation and everyday discursive practices' (Oslender 2004: 970). The river is more than a physical presence and a source of livelihood; because of these facts it also becomes bestowed with symbolic meaning, emotional significance, and political power. The river is more than just a river.

Source: information extracted from Oslender 2004

The value – and the need – of paying attention to these classes of people, types of actions and categories of events is that they constitute or make up daily life. For obvious reasons, a tsunami in the Indian Ocean, a hurricane in the southern coastal states of the United States, or a famine in the Sahel of Africa attracts and excites our attention. However, these are extra-ordinary, punctuating events that throw lives out of kilter. 'Normal' living is disturbed. The structures that govern 'ordinary' life are upturned – as was so bleakly evident in New Orleans in the aftermath of Hurricane Katrina in August and September 2005. But for the years before and after such events, whether they are natural or human-induced, the patterns and rhythms of life are tuned very differently. While, however, events like the tsunami may be traumatic and extraordinary, how and why people and communities respond as they do is intimately linked to local historical trajectories, local cultural norms, and local social structures (Box 1.2). Such events may be atypical but understanding their impacts and effects requires that the events are embedded in everyday geographies which, perhaps only for a short time, become particular day geographies.

In addition to the everyday being concerned with normal living rather than abnormal events, there is also a focus on 'ordinary' people. As is explored in greater detail in Chapter 8 (see page 167), the central motivation behind the Subaltern Studies project in India was a desire to avoid elitist historiographies and to acknowledge the important contribution of ordinary people on their own to the history of the subcontinent, independently of the elite (Guha 1982a: 3). Furthermore, as well as being 'everyday', a corollary of this is that the book also has a concern for personal geographies. Chambers (2004), in writing of the neglect of the personal dimension in development, says: 'It is self-evident to the point of embarrassment that most of what happens is the result of what sort of people we are, how we perceive realities, and what we do and do not do' (Chambers 2004: 12). So the focus on the everyday is not only because normal living is everyday living, but also because the everyday begins and ends with the personal.

BOX 1.2 The 2004 tsunami: the everyday effects of a global catastrophe

The Indian Ocean tsunami of 26 December 2004 was a global event and, on some measures, the greatest environmental disaster of the previous hundred years. More than 220,000 people were killed in 12 countries across South Asia, South-east Asia and East Africa. Something like 1.6 million people were displaced and it dominated news headlines across the globe for the first weeks of 2005.

It is easy when events are of such magnitude to simply become overwhelmed by the figures. But the tsunami provides a telling insight into why the local, the grounded and the everyday is so important if we are to understand and explain such events. Figure B1.1 sets out one feature of the tsunami: more females died than males. In some places this was at a ratio of 4:1. It is possible to guess at some of the reasons why this pattern occurred, but it is only through engaging with individuals and communities and their lived existences that the complex intersection of an array of factors becomes evident (Illustration B1.1).

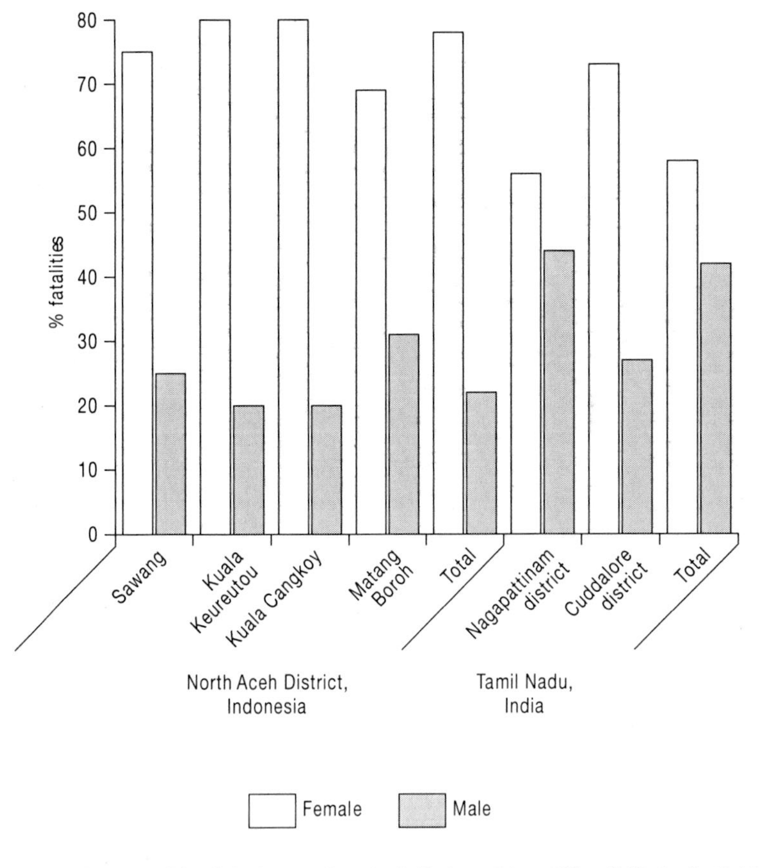

Figure B1.1 Pattern of fatalities by gender, Aceh (Indonesia) and Tamil Nadu (India) (2004)

Illustration B1.1 Ban Nam Khem near Khao Lak in southern Thailand. The village was virtually destroyed by the tsunami of 26 December 2004. By July 2005, when this picture was taken, the village had been rebuilt although many of the survivors were still too traumatised to 'return' to their new homes

To begin with the most obvious: men are, often, stronger and faster than women and were better able to escape or survive the wave. In addition, and less obviously, it seems that men were also more likely to be able to swim and were better at climbing. In many countries it is regarded as unseemly for women to be seen climbing trees and many would not have done so since childhood. Clothing in countries like Indonesia (Aceh is strictly Muslim) and Sri Lanka and India (where the sari is traditionally worn) may also have impeded women as they frantically struggled to escape the wave. It seems that patterns of employment in many of the afflicted coastal communities were such that women were far more likely to be killed. In fishing villages the men were often out at sea (the wave hit the coasts of Thailand and Indonesia in the morning); many scarcely noticed the wave. In coastal farming communities, men were working inland, on their fields. Women, by contrast, were far more likely to be at home engaged in home-based productive and reproductive tasks. When the wave washed ashore they were often encumbered as they tried to save their children. There were numerous reports of women trying to climb trees and hold onto branches while clasping a child under one arm, or trying to flee the wave with a young family in tow.

The 2004 tsunami offers an overarching story of natural hazard, risk, vulnerability and disaster. It also offers many, many community level and personal stories of human frailty and resilience, of social capital and official incompetence, of serendipity and sheer bad fortune.

Globalisation from below or grassroots globalisation

It was, perhaps, inevitable that scholars would, in time, react against the tenor of the ultra-globalists and begin to find something more complicated and contingent occurring. 'Globalisation from below' (Appadurai 1999, 2000; Taylor et al. 2001) and 'grassroots globalisation' are attempts to acknowledge – and promote – an alternative narrative, one that relishes in difference and recognises that the local can influence events, resist domination, and build alternative futures. The growing importance of localism is, in essence, a political expression of *either* the desire to resist and counter globalisation, or to rework globalisation to the benefit of local communities. This distinction is important. Those who hold to the 'resist and counter' position see globalisation as intrinsically and inevitably destructive and exploitative.[5] Those who embrace the 'rework' position see scope for making globalisation operate to the benefit of the Global South, local communities, and the environment. This is often based on treaty renegotiation and the reform and rebalancing of international institutions such as the World Trade Organization, the International Monetary Fund and the World Bank.

Neo-liberal globalisation is seen, by its critics, to have marginalised and pauperised peasants and workers. Its bulldozing logic has led to the 'race to the bottom' noted earlier, part of a process of 'competitive austerity' where countries compete to become the cheapest locations for global capital, to the ultimate detriment of those who are drawn in to the production process. Grassroots globalisation is a reaction to this and an attempt 'by marginalized groups and social movements at the local level to forge wider alliances at their growing exclusion from global neo-liberal economic decision-making' (Routledge 2003: 334). Both globalisation and the defensive response, grassroots globalisation, are composed of networks that operate across scales. This has led some scholars to see the network as more important than scale, whether that is global, regional, national and/or local.

The challenge – and therefore the difficulty – of scale for some human geographers is that scale, whether that is local, regional or global, is socially produced. This means that scale has no fixity, whether over historical time or between groups and individuals. It also means that because scale is not fixed it is also, and inevitably, contested (see Amin 2002). To put it simply: the *realm* and therefore also the *meaning* of the local for an international banker in London, a Ghanaian cocoa farmer, a German civil servant, a Mexican factory worker, an American stunt man, or a Filipino domestic worker in Hong Kong, is going to be very different for a range of reasons that span national context, living and employment patterns, language, age and gender, educational attainment, and so on. In light of this – and bearing in mind the increasing flows and fuzziness driven by globalisation processes – what and where, exactly, is 'the local'? Responding to these challenges, Amin makes a case for a 'non-scalar' interpretation of contemporary globalisation:

> My principal claim has been that the growing routinisation of global network practices – manifest through mobility and connectivity – signals a perforation of scalar and territorial forms of social organisation. This subverts . . . traditional spatial distinctions between the local as near, everyday, and 'ours', and the global as distant, institutionalised, or 'theirs'.
>
> (Amin 2002: 395)

So, for Amin, globalisation does not mean that one scale is becoming privileged over another (global over local) and, in the process, more powerful, but that the links, networks,

the outcome. Women were more mobile, but this was constrained and limited by what was deemed seemly. Women, and particularly those working in fair trade groups, were also more confident and willing to offer their opinions but they nonetheless kept out of village decision-making and local politics. In other words, while fair trade production such as that managed by Swajan can lead to positive change, 'empowerment' is sharply limited by the social and cultural context within which production is embedded. At the same time, of course, this context is evolving and one of the key issues is whether the pressure being placed on the boundaries of accepted practice by fair trade activities will, in time, lead to a more fundamental reworking of 'norms'.

Source: personal communication, Ann Le Mare

whereby the actions of humans play a role in producing and reproducing social structures, while social structures limit, constrain and enable human action (see Box 2.1). There is also a geography to Giddens' work in his concern to acknowledge the distinctiveness of the *locale* and the role of *locality* (as it became known among geographers) in providing the human and physical context for everyday living (see de Haan and Zoomers 2003: 351–2).

In a post-structuralist vein, Pierre Bourdieu developed the notion of *habitus* as another means of reconciling agency with structure, or the individual with society. Habitus is a set of dispositions, embodied in individuals, which generate certain practices and perceptions. These dispositions are inherited and are reproduced over time so that a regularity and a continuity emerges – a 'grammar' of living. However this regularity is not set in stone and is not rule driven so that there is scope for individualism and individuality. Undertaking research in Algeria during that country's war of national liberation in the 1960s, Bourdieu witnessed and was able to illuminate using ethnographic methods the collision of precapitalist norms with the economic rationality of colonisation. This 'mismatch' led Bourdieu to realise with 'total clarity' that 'access to the most elementary economic behaviours (working for a wage, saving, credit, birth control, etc.) is in no way axiomatic and the so-called "rational" economic agent is the product of quite particular historical conditions' (Bourdieu 2000: 18). For Bourdieu, then, universal 'givens' which mark out certain types of people as misfits and their activities as deviant are 'the product of a quite particular collective and individual history' (Bourdieu 2000: 28). There are, in short, patterns in the human landscape which the scholar should try to identify and explain but, at the same time, it is also possible 'to enter into the singularity of an object without renouncing the ambition of drawing out universal propositions' (Bourdieu 1986: xi). Actions are a product and a reflection of habitus, but habitus is shaping rather than determining.

How do Giddens' and Bourdieu's approaches to explanation work in practice? An example that implicitly draws on their work is Pieke's (2004) examination of political structures and agency in rural China – in this instance, in Xuanwei county in the north-eastern corner of Yunnan province, in the south-west of the country. Pieke is concerned to show how the state is not beyond and outside society at the local level, but is part of and produced by society. He explains that patterns of decision-making by cadres in Xuanwei can create a bureaucratic habitus of formidable force, 'privileging . . . certain options, while rendering others impossible or simply irrelevant' (Pieke 2004: 531). But when these cadres were removed following village committee elections in 2000, the habitus and the mould were broken. Social action is, in Pieke's study, an outcome of social structure, and constitutive of that structure.

If we permit not just a degree but an expectation of agency in most contexts then understanding everyday geographies becomes necessarily contingent. As Zoomers (1999) found in her study of the Bolivian Andes, notwithinstanding the structural components of life (geography, rural/urban location, agro-ecology, access to roads and markets . . .) there are many poor people who have become rich, and rich people who have become poor. On first sight, the situation in the Bolivian Andes appears unchanging and the population homogeneous. Closer inspection, though, reveals a much higher degree of dynamism and heterogeneity, so much so that the indentification of 'categories' becomes a fruitless exercise. I found much the same in my study of livelihoods in rural Laos:

> Reviewing the 55 case studies, and looking through the notes from the key informant interviews and group discussions, one of the most striking features was how far it was *normal* for households to buck the trend and deviate from the expected state of affairs.
>
> (Rigg 2005: 165, emphasis in original)

It is easy to think that the increasing prominence of qualitative methodologies in geography and the other social sciences, and the emphasis on life histories and life stories explored through ethnographic methodologies, has shifted the fulcrum towards an agency-oriented perspective (Illustration 2.1). However, it is not just *what* a researcher does in terms of research methods and approach, but *how* a researcher chooses to use the resulting data and information. When such personal geographies are reworked as representative of 'poor women's experiences' or 'small farmers' experiences' (i.e. are reified) then they become, in effect, about structure rather than about agency. They become emblematic of the manner and way of life of groups in society, those groups being defined in structural terms and categorised according to selected criteria. With this tendency in mind, Brettell argues that the 'goal [of life histories]

Illustration 2.1 Setting up a focus group discussion of rural women in Laos

is not to find a typical or representative individual but to assess how individuals interpret and understand their own lived experiences' (Brettell 2002: 439). This warning is well taken, but it is not easy to achieve. Categorisation is often a first step to understanding, and from here to explanation. So while researchers may not go into the field with the express intention of finding a typical or representative individual or household, having collected life histories it is all too easy to endow selected cases with just these qualities.

From structure and agency to livelihoods

The ongoing debate over structure and agency and how they can be reconciled has informed the growing concern for and interest in livelihoods. The structural-functionalist and *dependencia* approaches and perspectives of the 1970s and 1980s tended to present people, and particularly poor people in the Global South, as victims of structural constraints and limitations. A livelihoods perspective, as we will see, places people back at the centre of attention and explanation, endowing them with a degree of agency to struggle against, take advantage of, and resist or rework their political, economic, social and environmental milieu. Agency, actors and action become the watchwords of this shift in attention.

The importance of recentring attention on localities and populations is not just an alternative means of looking at and understanding livelihoods; it has also, directly and indirectly, informed a whole range of other developments and initiatives. In terms of research it can be seen reflected in the central importance of participatory methods; in politics, in the widespread and continuing efforts to empower local communities through decentralisation and devolution; and in economic terms, in the emphasis on community management of resources. In these ways, the focus on the local is not just another viewpoint – another way of seeing. The local becomes the source of knowledge and experience, the locus for decision-making, the context for management, and the site for action. Robert Chambers' (1997) book *Whose reality counts? Putting the first last* reflects this. Not only does the book make a passionate case for focusing first and foremost on the 'last' (the poor, the excluded and the disenfranchised), but also it stresses the extent to which research and action should begin at this level, so that the 'last' become subjects and agents and not merely objects guided from above:

> The problem is how, in conditions of continuous and accelerating change, to put people first and poor people first of all; how to enable sustainable well-being for all. . . . Basic to a new professionalism is the primacy of the personal.
>
> (Chambers 1997: 14)

Livelihoods and sustainable livelihoods

The interest in livelihoods, or sustainable livelihoods (SL), is usually traced back to the 1992 World Conference on Environment and Development and to a discussion paper by Robert Chambers and Gordon Conway also published in 1992 (Chambers and Conway 1992).[2] Given both these scholar-practitioners' long engagement with developing more people-centred methodologies (beginning with rapid rural appraisal (RRA) and, more latterly, participatory action and research (PAR)) and grounded approaches to scholarly research, particularly in poor, rural communities in the Global South, it is easy to understand why and how they arrived at developing the livelihoods perspective and approach (Table 2.2). In their 1992 discussion paper, Chambers and Conway offer the following definition of a livelihood:

> A livelihood comprises the capabilities, assets (including both material and social resources) and activities required for a means of living. A livelihood is sustainable which can cope with and recover from stress and shocks, maintain or enhance its capabilities and assets, and provide sustainable livelihood opportunities for the next generation; and which contributes net benefits to other livelihoods at the local and global levels and in the short and long term.
>
> (Chambers and Conway 1992: 7–8)

Ellis (1999), more succinctly, defines a livelihood as:

> The activities, the assets, and the access that jointly determine the living gained by an individual or household.
>
> (Ellis 1999: 2)

Figure 2.1 provides a schematic outline of the sustainable livelihoods (SL) approach which was drawn up to inform the 1997 UK Government White Paper on International Development. It has since been much copied and adapted by international and national organisations and NGOs (as well as academics) including the United Nations Development Programme (UNDP), World Bank, Food and Agricultural Organization (FAO), Consultative Group on International Agricultural Research (CGIAR), International Fund for Agricultural Development (IFAD), Swedish International Development Agency (SIDA), CARE International and the International Institute for Environment and Development (IIED). It is

Table 2.2 Livelihood studies: (selected) origins and development

Precursors and inspirations		*Development and related outcomes*
Methodological		***Methodological***
Rapid rural appraisal (RRA) (1980s)	⇨	Participatory poverty assessment (PPA) (1990s)
Participatory action and research (PAR) (1990s)		
Practical/applied perspectives		***Practical/applied perspectives***
Farming systems research (1980s)	⇨	DFID and Oxfam's sustainable livelihood approaches (1990s)
Agro-ecosystem studies (1980s)		Social development
Community development (1960s and 1970s)		World Bank's Social Capital Initiative
		Community participation and empowerment
Theoretical/conceptual		***Theoretical/conceptual***
Actor orientated research (Norman Long, 1980s)	⇨	Social capital (Robert Putnam, 1990s)
New household economics (Gary Becker, 1980s)		Social exclusion (1990s)
Entitlements (Amartya Sen, 1980s)		
Structuration theory (Anthony Giddens, 1980s)		
Habitus (Pierre Bourdieu, 1980s)		

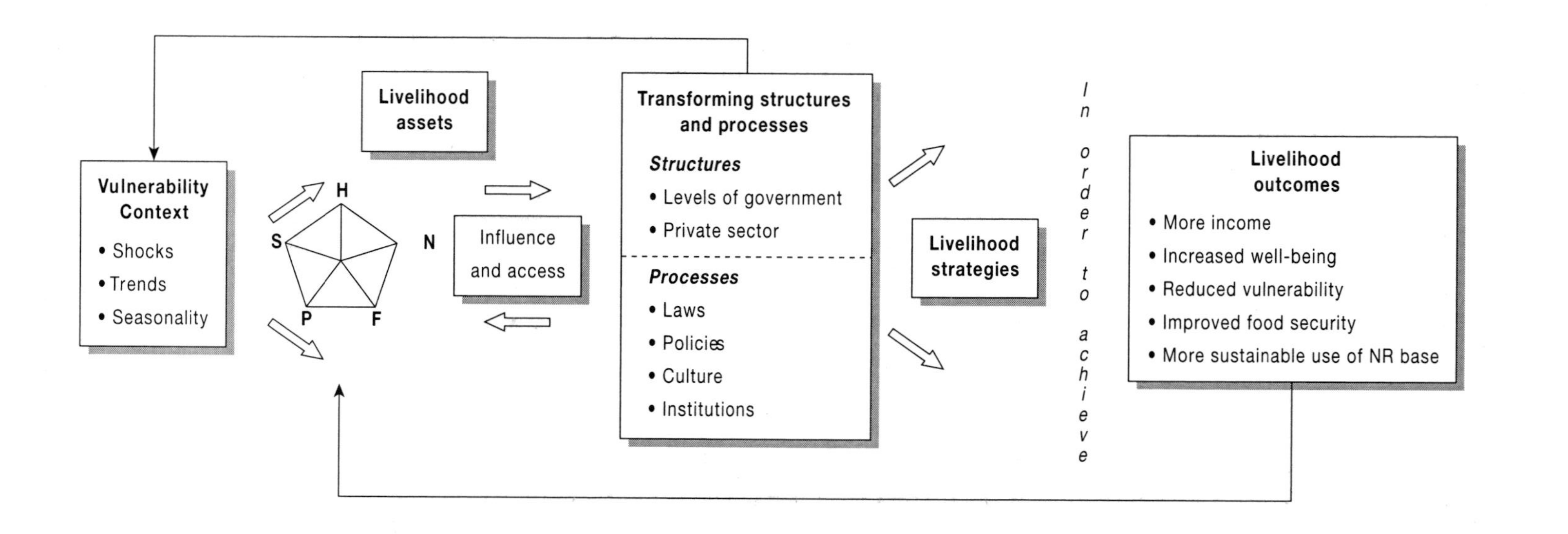

Figure 2.1 The DFID sustainable livelihoods framework

important to note that the SL approach and sustainable livelihoods analysis (SLA) are intended to be analytical frameworks to guide the identification of areas for development intervention in the context of combating poverty. They are not intended to reflect reality (Farrington et al. 1999).

Working through Figure 2.1 from right to left, households and individuals pursue certain livelihood outcomes through 'livelihood strategies'. These are filtered through political and institutional structures and draw on a pentagon of livelihood assets which are associated with five types of capital or livelihood building blocks – human, natural, physical, social and financial. How these assets are deployed is, in turn, linked to the types, frequency (periodicities) and intensities (severity) of vulnerabilities and shocks that households and individuals face. The figure gives the impression of a neat progression in terms of understanding but supporters of the SL approach have always been at pains to emphasise the fluid and shifting nature of livelihoods; nothing is neat and tidy. The core principles of SLA are:

- a focus on people and communities rather than on structures and the national context
- a concern with seeing livelihoods in holistic terms crossing sectors, spaces, actors and institutions
- and a commitment to identifying the macro–micro linkages that are salient to understanding livelihoods.

The key terms in the Chambers and Conway (1992) definition quoted above are: capabilities, assets, sustainable, stress and shock. Other important terms which have come to be associated with the sustainable livelihoods framework include: capital(s), coping, risk, resilience, vulnerability, security and well-being. Definitions of each are set out in Table 2.3.

There has been a tendency to view livelihoods in empirical and largely material terms: a livelihood is the way that an individual or a household 'gets by'. A livelihood is, therefore, about money, food, labour, employment and assets. The tendency to write about livelihood 'strategies' reflects this functionalist approach which may also be related to the way in which the framework was embraced by development agencies such as the UK's Department for International Development (DFID) and by the World Bank and UNDP. It is also reflected in Scoones' (1998) concentration on how actors mobilise various capitals and social resources in order to secure their livelihoods framed, in his paper, in terms of economic outcomes and planning interventions (see Arce 2003: 206). That SLA should have become framed in this way is not surprising in so far as the agencies, institutions and individuals who have taken the idea forward have seen its rationale lying primarily in its practical application.

For the purposes of this book, however, such a view of and approach to livelihoods is too mechanistic, too instrumentalist and too narrowly drawn in two key respects. To begin with, livelihoods are at least as much about the social and cultural bases of life and living, as the material ones. Therefore, why an individual should 'choose', for example, to work in a factory rather than in the field is often linked to such issues as aspirations, the status that is accorded to different ways of making a living, household family structures and patterns of authority, and networks of information and contacts. Without an understanding of the social or non-material foundations of work and living (which crucially contribute to the creation of well-being: see page 36) a very large slice of the explanatory fabric is obscured.[3] It should also be clear from these two sentences how Giddens' structuration and Bourdieu's habitus, discussed in the previous section, come to play a role in influencing the actions and activities of individuals and households. So, livelihoods need to be understood in structural context, if not in structural terms. They also, though, need be understood in *relational* terms: livelihoods

Table 2.3 Key terms and definitions

Assets	Material (physical) and non-material (social) resources that people possess
Capabilities	What people can do with their assets and entitlements
Capital	Livelihood resources consist of five forms of capital: natural, economic/financial, human, social and physical
Coping strategies	Coping strategies are strategies that individuals, households and communities embrace in response to a shock when normal strategies fail
Resilience	The ability of households or individuals to recover from shocks and avoid a decline in well-being
Risk	The likelihood of being faced with a decline in well-being/livelihood; risk exposure is the likelihood that a risk will occur
Security	The delivery of a 'reasonable' (locally determined) level of well-being to all people and at all times
Shock	A sudden, unpredictable and traumatic event which leads to a marked decline in well-being
Social capital	The 'glue' or 'fabric' of society including trust, behavioural norms, networks, contacts and connections
Stress	Small, predictable and, often, continuous and cumulative pressures
Sustainability	The ability of a natural or human system to maintain output and productivity
Vulnerability	The probability of being exposed to risk
Well-being	A normative concept encompassing both subjective (feelings, prestige, self-worth, freedom) and objective (income, housing, education) criteria: well-being is contextually determined with no fixed definition; some scholars prefer 'well-living'

Sources: Chambers and Conway 1992; Scoones 1998; World Bank 2001; Gasper 2004

are hewn out of the landscape through the ways in which different actors and their livelihoods interact and interrelate (see Murray 2002). At its simplest and starkest level, one person's wealth may beget the poverty of another.

The political ecology of cotton farming by smallholders in southern Mali nicely illustrates this wider point. Moseley (2005) sampled 133 smallholders in Siwaa and Djitoumou, stratifying these households by wealth (Figure 2.2). Relatively richer farmers, as a group, were more deeply involved in cotton production, with more than three times as much land planted to the crop (6.2 versus 2.0 hectares). Because these richer farmers could afford to prepare their land using plough and oxen and, on occasion, tractors they were able to extensify their cultivation onto more marginal, village lands. This land, part of the village stock, was more prone to erosion and degradation and its use by wealthier farmers left less available for alternative uses. In this way the costs of extensifying production were borne by the community as a whole rather than by the richer households who benefited from its cultivation. In a similar vein, richer households had more cattle. These were grazed on common village lands which were also used, particularly by the poor, for shea nut collection.[4] Once again, richer households derived disproportionate benefit from the use of common lands (because they grazed more cattle) while the poor suffered disproportionately as the effects of over-grazing diminished the productivity of the community resource, undermined shea nut production,

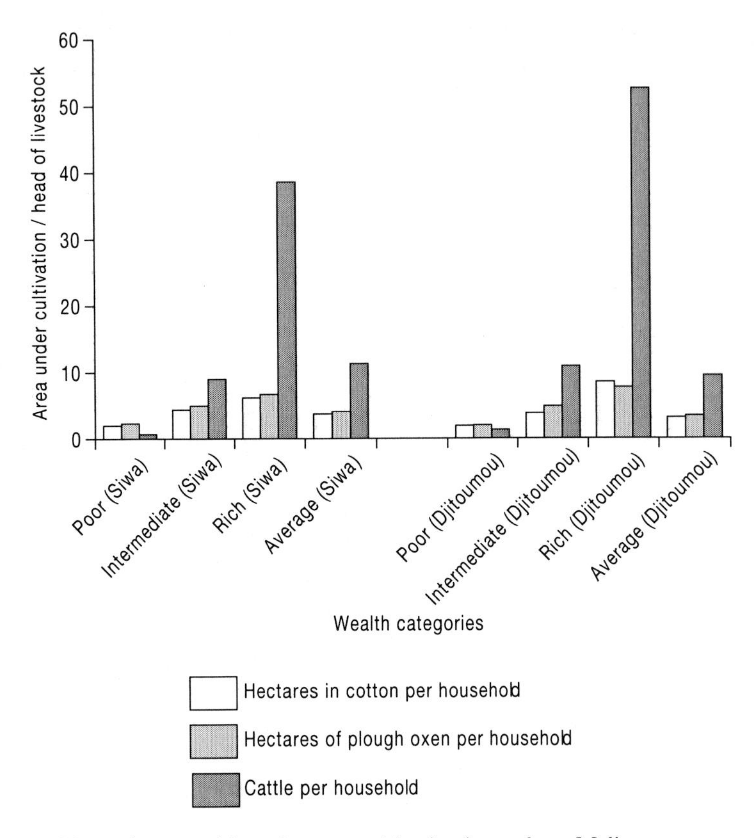

Figure 2.2 Wealth, cattle ownership and cotton cultivation in southern Mali

and put pressure on fallow reserves. This case from Mali, therefore, clearly shows how livelihoods, though they may be enacted (and conceptualised) at the level of the individual or household, include sets of relationships and trade-offs which require a consideration of the individual or household and their relational place in wider society.

The second shortcoming of seeing livelihoods in material terms is the tendency to overlook the dynamism inherent in patterns of living and their evolution over time. This may be linked to seasonal fluctuations, life cycle changes as households make the transition from 'young' to 'mature', national economic trajectories and the changing opportunities afforded by economic progress, or whether a household is upwardly or downwardly mobile. While the concern for identifying and supporting 'sustainable' livelihoods may give the impression that the temporal dimension is central to the concept, there is a fair degree of muddle when it comes to clarifying exactly how sustainability is defined and measured (Scoones 1998: 5; Murray 2002). Furthermore, the emphasis is on how to sustain livelihoods and this overlooks the degree to which, in the context of sometimes rapid economic development and deep cultural change, livelihoods are being continually reworked and transformed over time. In other words, we need to appreciate not just how households and individuals sustain their existing patterns of living – often reflected in the tendency to write in favourable terms of livelihood 'resilience' – but how (and why) the bases of livelihoods change over time.

These two shortcomings, as de Haan and Zoomers (2005: 33) observe, are also two of the key challenges facing livelihood studies. First, how do you get a sense of the non-material propelling and moulding forces? It is easy enough to measure and calibrate such things as land, physical assets and income, but what of preferences, expectations and aspirations? These are much harder to assess but are just as important in explaining the decisions that people take and the choices they make. The second challenge is to imbue our understanding of livelihoods with a sense of dynamism which is not just an 'add-on', but inherent and intrinsic to the approach. There is often a romantic undercurrent to livelihood studies which seems bent on preserving the status quo on the assumption that a sustainable traditional livelihood is necessarily better than a new one. For disenfranchised Latin American farmers this may be true; but what of Malaysian factory and office workers who have abandoned their former rural lives and agricultural livelihoods for urban living?

There is, also, a third often-commented on shortcoming with livelihood studies: the neglect of politics and power relations. Even if one were to gain a full understanding of the material and non-material bases of livelihoods, and the full range of assets and endowments that an individual or household could – in theory – access, it is often the case that there is a gap or disjuncture between endowments and assets (capitals), and entitlements, and between entitlements and capabilities. This is because assets and endowments are not there for the taking, but are selectively accessed or awarded according to prevailing systems of political and social relations. This is one of the themes developed, for example, in Twyman et al.'s (2004) study of livelihoods among small holders on the margins of the Kalahari in southern Africa (Botswana and North West Province, South Africa) where, they conclude, neither wealth or poverty are the key to securing (and understanding) livelihoods. Rather, sustainable livelihoods are founded on the ability of households to *mobilise* the assets that they have. Thus, in Table 2.4, a young woman may have energy, skills, youth and initiative but because of the nature of the operation of the household and wider society she will find that returns to her work are limited, upward mobility stymied, and the income that she does generate largely syphoned off by the head of the household.

Livelihood 'pathways'

It is possible to see the debates over structure, agency and livelihoods being brought together in the suggestion that we should focus on identifying and understanding livelihood 'pathways', rather than livelihoods per se:

> A pathway can be defined as a pattern of livelihood activities which emerges from a co-ordination process among actors, arising from individual strategic behaviour embedded both in a historical repertoire and in social differentiation, including power relations and institutional processes, both of which play a role in subsequent decision-making.
>
> (de Haan and Zoomers 2005: 45)

The attraction of viewing livelihoods in terms of pathways is that it is both dynamic and also permits a degree of agency. Even so the term 'pathways' is rather too constricting. Experience shows that people 'jump' pathways: serendipity and simple bad luck can cause livelihoods to be reworked in such a way that the latitude offered by the notion of a pathway is simply insufficient to accommodate the degree of change that can arise.

Table 2.4 From assets and endowments to entitlements and capabilities

Assets, capitals and endowments	⇨	*Power relations/politics*	⇨	*Entitlements*	⇨	*Capabilities*
From assets to capabilities: a generic guide						
The rights and resources that people (individuals or groups) have	⇨	Power relations and social context	⇨	Commodities over which people have command	⇨	What people can do with their entitlements
Example 1: Female Marantaceae *leaf-collector, Ghana*						
Marantaceae-rich forests; village and household membership; labour	⇨	Access rights to village, forest reserve and household lands; conflicts settled by village appointed 'queen mother'	⇨	Access rights to forest reserve gained through permits; to village lands through membership of village; to household lands through household membership; time to collect leaves negotiated with husband and co-wives	⇨	Control over income generated from collection and sale of leaves negotiated with husband in context of household needs
Example 2: Factory daughter						
Skills, youth, energy and initiative	⇨	Constrained by household norms, working in a patriarchal factory context	⇨	Patriarchal factory environment limits upward mobility, limits female agency and segments opportunities by age and gender	⇨	Salary is claimed by household head; use to which income is put dictated by parents

Sources: Example 1 is taken from Leach et al. 1999; Example 2 is generalised.

Wealth and well-being

Most studies of the Global South with a development orientation focus on income, prosperity, wealth and poverty. The development game has been one where the object, and the measure, of success is the achievement of economic growth, the raising of incomes, the increase in prosperity, and the decrease of poverty. Wealth and poverty, however, have certain cultural connotations and some scholars prefer to think, instead, of 'well-being' (and, by association, 'ill-being'). As Chambers (2004: 10) says, 'extreme poverty and ill-being go together, but the link between wealth and well-being is weak or even negative: reducing poverty usually diminishes ill-being; amassing wealth does not assure well-being and may diminish it'.[5] A second advantage of thinking of well-being rather than poverty and prosperity is that it distances the understanding of the circumstances of living from development, as both a process and a project. Finally, thinking in terms of well-being permits non-material aspects of livelihoods to be accorded a centrality which is generally absent from poverty studies. Changing values may lead to a decline or an increase in well-being even without any material change in the livelihoods profile (in terms of income, diet or possessions) of an individual or household.

A project based at the University of Bath in the UK has attempted to disentangle some of the elements that comprise 'well-being' through a comparative study of four countries: Bangladesh, Ethiopia, Peru and Thailand.[6] At the heart of this project is the notion, articulated in the previous paragraph, that 'development', in material terms, does not automatically translate into well-being:

> Wellbeing must combine the 'objective' circumstances of a person and their 'subjective' perception of their condition. Furthermore, wellbeing cannot be thought of only as an outcome, but as a state of being that arises from the dynamic interplay of outcomes and processes
>
> (McGregor, 2007)

The work on Bangladesh (Camfield et al. 2006), for example, highlights the apparent anomaly that while Bangladeshis are some of the poorest (in material terms) people in the world, they are also among the happiest. Furthermore, at a time when poverty has been declining in Bangladesh, and the country's human development index rising, happiness has been falling. Drawing on fieldwork undertaken in six sites across the country (two rural, two peri-urban and two urban), the project has attempted to make some sense of these counter-intuitive data. The first point is that income and material well-being do matter in Bangladesh and those who are well paid and in full-time employment tend to be happier than those who struggle to find work and get by (Figure 2.3). But this is complicated by the presence of many other indicators and determinants of well-being. In particular, it is the quality of relationships – with family and community – which appears to be of central importance. This, though, is not separate from material well-being, because the quality of relationships can positively enhance incomes, and vice versa.

> Our findings highlight the centrality of relationships to people's subjective wellbeing. . . . Relationships determine individuals' values, choices, actions, and indeed the construction of self. More than any other factor, they determine what people are able to do or be, and what they actually achieve or become.
>
> (Camfield et al. 2006: 25–26)

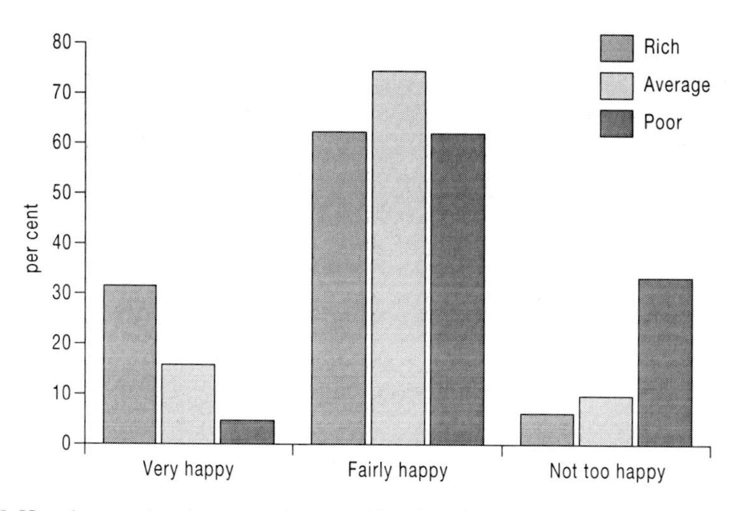

Figure 2.3 Happiness and socio-economic status, Bangladesh

Livelihoods, scale and the roots of change

The opening chapter to this book discussed the challenge of scale, particularly in connection with the global–local debate. Scale is relevant here too. As noted above, the livelihoods approach arose, in part, because of a desire to escape from the perceived straitjacket of structural perspectives, in an effort to re-emphasise locality, and to re-energise the individual. The danger, though, is clear. The livelihoods approach, in drilling down to the local and the personal, stands the risk not only of reifying community but, at the same time, of ignoring processes and the structures of life which lie above the local but which impinge on the local.

In theory this should not happen because by focusing on livelihoods, rather than on situations, then if people are engaged in multi-scalar ways of living, or if local systems are implicated in wider circuits of production, the livelihoods framework should lead on to these higher and wider levels of analysis. Understandably – and perhaps inevitably – though, the experience of much livelihoods research has been that it has focused on the capacities of local actors, on the assets available at the local level, and on the nature of local level institutions. This not only undermines the claim that the sustainable livelihoods approach is holistic but also overlooks the propelling forces behind many of the key changes that are evident across the Global South. These are, *inter alia*: diversification, deagrarianisation, delocalisation, differentiation and restructuring. Taken together, these processes are playing an instrumental role in changing livelihoods in the Global South, and their origins lie not in local level resources, whether those are seen in terms of physical assets, traditional wisdoms or local technologies, but in developments occurring above and beyond the local stage.

In writing this I appreciate that it might seem that I am drawing a distinction between the 'in here' and the 'out there' (to use Ash Amin's distinction). To be sure, and notwithstanding all the scholarly talk of nested scales and glocalisation (see Chapter 1), there *is* a division which is much more than imagined. When livelihoods become implicated in wider scales and levels, whether that is through, for example, contract farming or heightened levels of mobility, it does bring some important qualitative changes to people's lives and livelihoods. The division is not necessarily a sharp or an absolute one, but it does matter.

The importance of not completely side-lining the role of scale on the basis that all is nested and interrelated can be seen in how notions of, and approaches to, participation vary in the operation of Save the Children in Burkina Faso in West Africa (Michener 1998). Depending on where one looks and who one asks, participation varies in this one organisation from participation as empowerment at the national level, to participation as 'commitment' in some of the manuals, to participation as a 'duty' and an 'obligation' among fieldworkers at the local level (Figure 2.4). That fieldworkers should take this view and approach is not altogether surprising because local people view participation as a means to extract resources from donors. This needs to be understood as an outcome of the historical engagement of people in Michener's research area with external agents. And this, in turn, informs and colours the ways in which local people respond to participatory development interventions: 'For [local people in Bazega] participation has little to do with self-reliance, empowerment or even efficiency. Instead, it is an opportunity to extract resources from willing agencies' (Michener 1998: 2116).

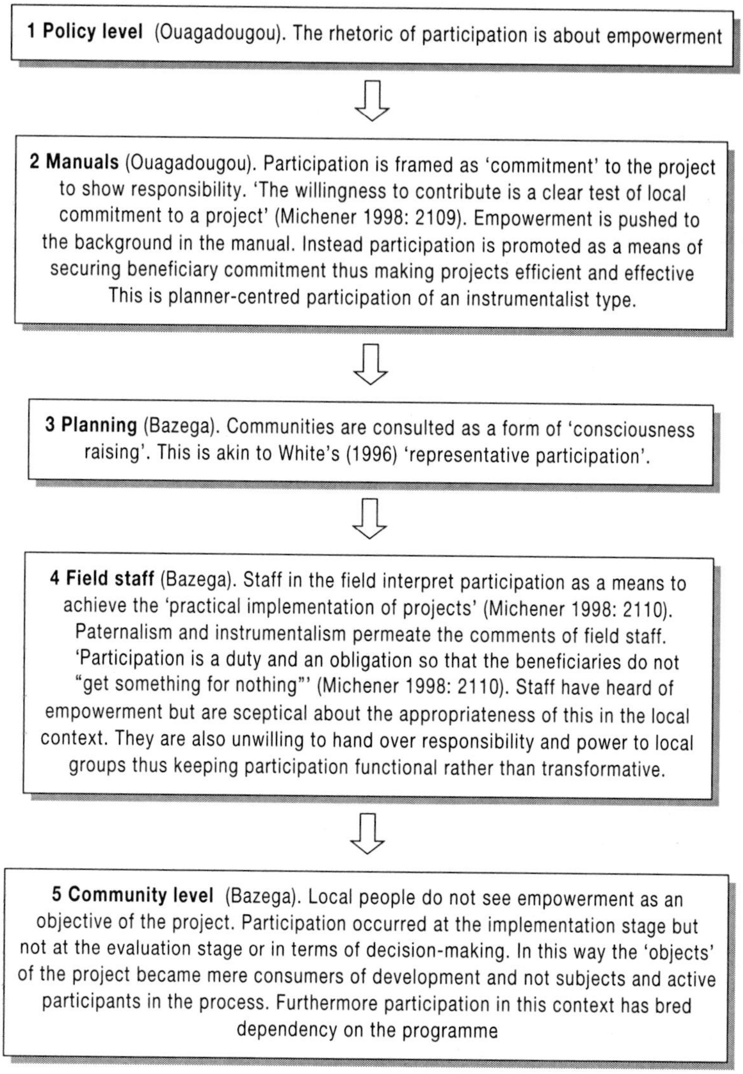

1 Policy level (Ouagadougou). The rhetoric of participation is about empowerment

2 Manuals (Ouagadougou). Participation is framed as 'commitment' to the project to show responsibility. 'The willingness to contribute is a clear test of local commitment to a project' (Michener 1998: 2109). Empowerment is pushed to the background in the manual. Instead participation is promoted as a means of securing beneficiary commitment thus making projects efficient and effective This is planner-centred participation of an instrumentalist type.

3 Planning (Bazega). Communities are consulted as a form of 'consciousness raising'. This is akin to White's (1996) 'representative participation'.

4 Field staff (Bazega). Staff in the field interpret participation as a means to achieve the 'practical implementation of projects' (Michener 1998: 2110). Paternalism and instrumentalism permeate the comments of field staff. 'Participation is a duty and an obligation so that the beneficiaries do not "get something for nothing"' (Michener 1998: 2110). Staff have heard of empowerment but are sceptical about the appropriateness of this in the local context. They are also unwilling to hand over responsibility and power to local groups thus keeping participation functional rather than transformative.

5 Community level (Bazega). Local people do not see empowerment as an objective of the project. Participation occurred at the implementation stage but not at the evaluation stage or in terms of decision-making. In this way the 'objects' of the project became mere consumers of development and not subjects and active participants in the process. Furthermore participation in this context has bred dependency on the programme

Figure 2.4 Hierarchies and levels of participation in Burkino Faso

Operationalising agency and livelihoods: case studies from Ghana and South Africa

So far the discussion has been mostly generic, divorced from people and place. The question, then, is how is such a livelihoods perspective operationalised in practice? What additionality, in practical terms, does it bring to our understanding of geographies of the Global South? This is best exemplified and explored through taking two case studies, one from Ghana and the second from South Africa. They could, equally, have come from a country in Latin America or Asia or, for that matter, from Europe or North America.

Women, Marantaceae *and livelihoods in Brong Ahafo, Ghana*

Leach et al. (1999) provide a case study of a livelihoods and entitlements approach to the use of *Marantaceae* in two villages in Wenchi District, Brong Ahafo region, Ghana.[7] This plant is particularly important for women in Ghana's forest zone where the leaves are sold and used to wrap food and other products. Access to the leaves becomes an endowment over which women gain rights. Some *Marantaceae* are found in government protected reserves; others are located on village commons; and still more trees grow on farm land. Collection rights in reserve forest are mediated through permits issued by the Forest Department. On village lands, collection rights require village membership, while on farm land they are dependent on belonging to the family that holds the land. But it is not simply a case of belonging to the village or to the land-holding family, or of negotiating a permit with the Forest Department. When they collect *Marantaceae* leaves women must negotiate with their husbands and co-wives for time because collection takes women away from other household tasks. There is, therefore, an opportunity cost at the household level. Furthermore, the income generated does not automatically accrue to the woman collecting and then selling the leaves; this must also be negotiated with the husband in the context of the wider household and its needs. Women collectors aim to secure income for their own use, so that they can support themselves and their children, but what amount of time they are allocated (or allowed) and what proportion of the income generated accrues to the individual is filtered through the institutions and structures of the household (particularly) and the village. The complexity of *Marantaceae* leaf collection does not end here. There are frequent conflicts between individual collectors and collection groups and these are mediated through a 'queen mother of leaf gatherers' who is appointed by each village. Finally, it is important to see this system not set in stone but continually adapting in the context of wider environmental, economic and social changes.

It is possible to see in the disturbed forests and groves of Ghana an example of how structure and agency interweave (see Table 2.4). As Leach et al. (1999: 239) write: 'the environment provides a setting for social action but is also a product of such action'. Access to *Marantaceae* leaves is negotiated at the levels of the household, the village and the local state (in the guise of the Forest Department). Access rights are embodied in formal, informal and communal structures. These are the structures through which women access (or do not) the *Marantaceae* resource and, in turn, the income generated from its sale. At the same time, though, people have the ability to transform the environment and therefore the resource through changing cropping patterns, altering levels of resource exploitation, or adjusting the length of fallow periods. 'Over time, the course of environmental change may be strongly influenced by particular conjunctures of institutional conditions, or by the coming together of contingent events and actions' (Leach et al. 1999: 239). In the case of these two villages in Brong Ahafo in Ghana the coincidence of structure and agency has resulted in biodiversity-rich and expanding forest patches in otherwise heavily settled areas, challenging the normal narrative of forest and environmental decline.

Differentiation and diversification in Qwaqwa, South Africa

Qwaqwa is a small 'homeland' in the Orange Free State of South Africa. Slater's (2002) work, which is part of a wider project examining agrarian change in the area, draws upon interviews with 125 households and the life histories of 45 individual respondents. 'By using the vivid testimonies of three life-history respondents', Slater (2002: 601) writes, 'we can find evidence for the patterns of differentiation that were common to many people in Qwaqwa, but simultaneously retain some sense of the uniqueness of ordinary people's lived experiences'.

The wider, structural and historical contexts which mould but do not govern livelihoods in Qwaqwa are on the record. They involve: the displacement of thousands of people from the towns and white-owned farms of the east Orange Free State in the 1970s and 1980s and their movement into Qwaqwa; the emergence of a well-connected bourgeoisie from the 1970s who were able to exploit their links with the politically powerful local chiefs as Qwaqwa became a self-governing 'homeland'; the growing availability of jobs in industrial parks in Qwaqwa from the 1980s as the Regional Industrial Development Programme stimulated employment growth in the homelands as a way of stemming the drift of workers back to South Africa; and, from the 1990s, the process of deindustrialisation and informalisation which followed post-Apartheid independence and which hit some of those who had accumulated wealth in the 1980s but had failed to diversify.

Slater then uses the experiences of Tshepe Kganya, a policeman and trader, Hadiyo, a labourer and door-to-door saleswoman, and Mmalefu Mokgatla, an urban worker, to illustrate the ways in which circumstance, serendipity, social capital and inter- and intra-household relations inject a degree of contingency into the structural frame outlined in the last paragraph. Tshepe Kganya's house in Soweto was burnt down by students in 1976 when Afrikaans was imposed as the medium of instruction, leading to the Soweto Uprising of June. This precipitated Tshepe Kganya's return to Thibella, a remote Qwaqwa village, where he set up a taxi business and established a chain of tuckshops. Hadiyo was twice 'stolen' from her school near Weltevreden by a man she didn't even like. Forced to marry him, Hadiyo then moved to her husband's home, where she set up a business selling milk and chickens. Mmalefu Mokgatla's adult life, from the time of her marriage at the age of 16, has been punctuated by forced removals, retrenchment and death, taking her, over three decades, from place to place and job to job, but with little sense of upward mobility. In October 1998 she was forced to sell three cast iron cooking pots, turn to the generosity of her neighbours, and grow crops in her back garden. Mmalefu Mokgatla's hopes and future rested on her son, Kete, then studying at UniQua.

Theorising-up from Qwaqwa and Brong Ahafo: three lessons, two warnings and a conclusion

What, in wider terms, do these two case studies from West and South Africa, and the earlier discussion of participation in Burkina Faso indicate? They provide, it is suggested, three lessons and two warnings.

The first of the lessons is that no matter what level of detail and understanding is acquired regarding a household's or an individual's livelihood, the 'circumspective' needs to be informed by the 'retrospective'.[8] In other words, the here-and-now of livelihoods (the circumspective) becomes truly meaningful only if it is informed by an appreciation of the historical circumstances and events that preceded it (the retrospective). The second lesson is that the structure–agency dichotomy is a shifting one where, and notwithstanding the problems with the distinction itself, structural elements intrude into everyday living at differing levels of intensity over time, and therefore offer differing levels of explanatory power depending on the moment of history being assessed. The third lesson, and this comes through more clearly in the Brong Ahafo and Burkina Faso examples, is the need not only to see livelihoods arising from historical processes but also to view them in relational terms. These relations may be between the members of a household (as in inter-gender and inter-generational relations), between households, or between other units at higher levels.

The first of the two warnings is that extrapolating from the retrospective and circumspective to the prospective – in essence drawing a line to the future – is necessarily risky and implies a level of certainty about livelihood trajectories which is often not borne out by experience. The defining events in the lives of Slater's three Qwaqwa respondents did not arise sequentially from what had gone before. They represented 'breaks' in the historical and livelihood narrative which could not be predicted. Yet it was these unpredictable events which played a large part in explaining current (or circumspective) livelihood conditions. The second warning is that livelihoods are organic, shifting, fluid and contingent. They are, therefore, not easily grasped (see Brocklesby and Fisher 2003). The rather managerialist approach of the DFID and World Bank to livelihoods plays down this contingency in a desire to identify livelihood types which can then be targetted for development intervention.

This chapter has taken a constructively critical view of the livelihoods approach. Despite its limitations the reasons why the approach was embraced by so many with such alacrity and enthusiasm remain pertinent: principally, the need to understand ways of life and living which embrace the complexity of factors at work. 'Holistic' is a term which is used rather too much and has, therefore, perhaps lost its edge and some of its meaning. However if we are to understand everyday living then the scope of view must extend from the cultural to the economic, from the social to the political, from the present to the past, and from the local to the global. This is, clearly, asking a very great deal indeed.[9]

Further reading

Two seminal publications on the structure–agency debate are Giddens (1984) and Bourdieu (2000). Brettell (2002) provides an overarching history. Livelihoods analysis is closely associated with the work of Robert Chambers (1988) and Chambers and Conway (1992); Scoones (1998) provides a good summary, Ellis (1999) a reflection on policy implications, while the work of de Haan and Zoomers (2003, 2005) is constructively critical. For case studies of livelihoods analysis being used in practice see the collection of articles in the *Journal of Southern African Studies* (2002, volume 28, number 3), as well as Zoomers on highland Bolivia. The shift towards participatory poverty assessments can be tracked through the World Bank's website and, in particular, their large collection of country-level PPAs from the 'voices of the poor' initiative (see: http://www1.worldbank.org/prem/poverty/voices/index.htm); the 2000/2001 World Development Report 'attacking poverty' is a helpful synthesis of the thinking and the results; an overview is downloadable from http://siteresources.worldbank.org/INTPOVERTY/Resources/WDR/overview.pdf. For various downloadable studies of well-being see the Well-being in Developing Countries website at http://www.bath.ac.uk/econ-dev/wellbeing/.

Structure, agency and structuration

Bourdieu, Pierre (2000) 'Making the economic habitus: Algerian workers revisited', *Ethnography* 1(1): 17–41.

Brettell, Caroline B. (2002) 'The individual/agent and culture/structure in the history of the social sciences', *Social Science History* 26(3): 429–445.

Giddens, Anthony (1984) *The constitution of society: outline of the theory of structuration*, Cambridge: Polity Press.

Livelihoods analysis

Chambers, Robert (1988) *Sustainable livelihoods, environment and development: putting poor rural people first*, IDS Discussion Paper 240, Brighton, Sussex: Institute of Development Studies. Downloadable from http://www.ids.ac.uk/ids/bookshop/wp.html.

Chambers, Robert and Conway, Gordon (1992) *Sustainable rural livelihoods: practical concepts for the 21st century*, IDS Discussion Paper 296, Brighton, Sussex: Institute of Development Studies. Downloadable from http://www.ids.ac.uk/ids/bookshop/dp/dp296.pdf.

De Haan, Leo and Zoomers, Annelies (2003) 'Development geography at the crossroads of livelihood and globalisation', *Tijdschrift voor Economische en Sociale Geografie* 94(3): 350–362.

De Haan, Leo and Zoomers, Annelies (2005) 'Exploring the frontiers of livelihood research', *Development and Change* 36(1): 27–47.

Ellis, Frank (1999) 'Rural livelihood diversity in developing countries: evidence and policy implications', *Natural Resource Perspectives no. 40*, London: Overseas Development Institute.

Scoones, Ian (1998) *Sustainable rural livelihoods: a framework for analysis*, IDS Working Paper 72, Brighton, Sussex: Institute of Development Studies. Downloadable from http://www.ids.ac.uk/ids/bookshop/wp.html.

Livelihood case studies

Journal of Southern African Studies (2002) 28(3): special issue on livelihoods in southern Africa.

Zoomers, Annelies (1999) 'Livelihood strategies and development interventions in the Southern Andes of Bolivia: contrasting views on development', *Cuadernos del Cedla no. 3*. Downloadable from http://www.cedla.uva.nl/10_about/PDF_files_about/boliviacderno.pdf.

Poverty and well-being

Well-being in Developing Countries website: http://www.bath.ac.uk/econ-dev/wellbeing/.

World Bank (2001) *World development report 2000/2001: attacking poverty*, New York: Oxford University Press. Overview downloadable from http://siteresources.worldbank.org/INTPOVERTY/Resources/WDR/overview.pdf.

3 Lifestyles and life courses

The structures and rhythms of everyday life

Introduction

It has already been said, more than once, that this book is not about development. But it is about change. It has also been noted that the book is not a structural interpretation of the Global South. However, it does take cognisance of the structuring contexts within which people live. This chapter fleshes out these two areas. First, it seeks to unravel the social and cultural frameworks that contribute to the structures of life. Second, it aims to illuminate modernity and the process of modernisation and how this is changing lifestyles and the expectations and aspirations that inform people's actions.

The chapter consists of three main sections. To begin with, it outlines the key structures of society and, in particular, the nature of the household and the community and the interrelationships and interactions that link – and separate – individuals, families/households, and communities. Then, it links these with life courses and the rhythms of life. Finally, the chapter sets out to highlight the ways in which modernity, broadly drawn, is impacting on these structures and rhythms of living.

The structures of everyday life

The household

More often than not, and this applies particularly to work on the Global South, the basic unit of survey and analysis in almost all social science is the household (Illustration 3.1). The household becomes, in effect, a single, welfare maximising decision-making unit so that, as Russell says of her work in Swaziland, 'it is household labour on household fields for household consumption' (1993: 756). The household becomes, de facto, an 'individual by another name' (Folbre 1986a: 20; see also Folbre 1986b). While households may exist in some sense or other in all societies, it is also clear that they exist differently. Is the household a site of production? Or reproduction? Does it have political significance? Is it spatially situated? And so on. But, there is one thing we can be sure of: households are collections of individuals (Crehan 1997: 94). And it is the analytically difficult relationship between the individual and the household which explains why few researchers dig any deeper, at least in terms of their units of analysis. It is extraordinarily difficult to extract the individual from the family and the household of which they are a component part. Not only is it difficult; such an effort would also be misguided. Individual desires, motivations and decisions are tempered, guided and sometimes dictated by the individual's locus within a household. Writing

Illustration 3.1 Pito (a traditional, cereal-based alcoholic drink) brewing outside a compound house in Ghana inhabited by five households. In four of the households the women participate in the brewing, sharing equipment and working together before dividing the beverage and walking around their respective patches selling it. They keep the money that they earn for themselves or their own household. In Ghana, men and women within the same household keep separate economies with responsibility for different expenses. Photograph: Kate Gough

of women's mobility, Chant (1998: 12) argues that 'even in instances where women ostensibly make their own decisions to migrate, it is hard to abstract household conditions from the process'. Households, she writes, 'not only [create] the material conditions for gender-selective migration but also [act] as filters for familial gender ideologies which impact upon motives for migration and the relative autonomy of migrant decision-making' (Chant 1998: 12).

In treating the household as a single, welfare maximising decision-making unit, however, scholars have traditionally glossed over the frictions, contradictions and inequalities that are inherent in the operation of the household. The (male) household head and decision-maker is 'portrayed as a benevolent dictator . . . who has internalized family members' needs, makes decisions with the collective good in mind, and rules with justice' (Wolf 1992: 15). Since the 1980s, however, scholars, and particularly those interested in gender relations and inequalities, have exposed the weaknesses in this Utopian vision of the operation of the household. They have sought to 'problematise' the household in the sense that they have exposed the fault lines that run through it. The key fault lines usually relate to relations between men and women and between the generations. Guyer (1981) writes:

> With a methodology based on [the] household as a major analytical concept, one cannot look at three critical factors, all of which seem to be changing in Africa today, with very important consequences: the relationship between old and younger men; the relationship

> between men and women; and the relationships among domestic groups in situations where wealth or control of resources vary widely.
>
> (Guyer 1981: 99)

These fault lines operate in small ways and large.

One of the more striking fault lines was the observation by the Nobel prize-winning economist Amartya Sen, some years ago, that '100 million women are missing' (Sen 1990a).[1] By saying this, Sen was highlighting the fact that given the biology of birth and survival there should have been, at the time he was writing, 100 million more women in the world than there actually were. Fewer girls were being born than should be expected,[2] and fewer of these were surviving:

> In view of the enormity of the problems of women's survival in large parts of Asia and Africa, it is surprising that these disadvantages have received such inadequate attention. The numbers of 'missing women' in relation to the numbers that could be expected if men and women received similar care in health, medicine, and nutrition, are remarkably large. A great many more than a hundred million women are simply not there because women are neglected compared with men. . . . We confront here what is clearly one of the more momentous, and neglected, problems facing the world today.
>
> (Sen 1990a)

While Sen makes a wider point about the general position of women relative to men, he is at pains to highlight two issues. First, the stark differences in the ratio of women to men that exist between regions, between countries, and within countries; and, second, the fact that there are marked differences even between countries with the same level of development. In other words this is not a simple issue of West versus non-West, or of rich versus poor, although both culture and wealth do play their parts. The explanation lies in a complex and shifting coincidence of social, economic, political and cultural factors. As Table 3.1 shows, Africa – the world's poorest continent – does rather better than South Asia in terms of its sex ratio. And within India, one of the country's poorest states – Kerala – has a more equal sex ratio than does the Punjab, India's richest state.

Sen's striking observation is underpinned by many smaller examples of discrimination which often contribute, brick by brick, to the bigger picture. Take, for example, the apparently simple question of who, in a family, chooses where a house should be situated. In the squatter settlement of Samé in Bamako in Mali, men choose the site for the location of the family home and compound. As Simard and De Koninck (2001) show, the criteria used for selecting a site are male-centred and such criteria 'do not necessarily coincide with those of the other family members, particularly the women, even though it is they [women] who use the compound most intensively' (2001: 30–31). The mother of one male compound owner acknowledged her lack of power to influence her son: 'Now [that] my son provides for me, I can no longer make decisions' (2001: 31).

It may seem, from this discussion, that the household as a unit of analysis and as a collective enterprise, is dead. Rather, however, it emphasises the need to look beneath the surface and to acknowledge that behind the walls of domesticity is a stage where cooperation and conflict, corporatism and individualism, mutuality and inequality, and consensus and discordance, coexist (Rigg 2003: 245). The household could be said to be defined by dissonance. Sen, once again, has explored this issue in his work on 'cooperative conflicts' (Sen 1990b). Conflicts are 'resolved' through established and 'agreed' patterns of behaviour. These patterns do not,

There has been a tendency to equate modernisation with Westernisation: by becoming more modern populations are also, and necessarily, becoming more Western in the sense that they are embracing values and attitudes that are 'Western' in origin (Ingham 1993: 1807). This slightly lazy coupling of the two processes (modernisation and Westernisation) is problematic because while the experience of many societies is that there is usually a link – sometimes a tight one and sometimes a loose one – between modernisation and a set of social and cultural changes that would seem, on first sight, to be akin to Westernisation, in detail the nature of modernity is shaped by the specific social and historical contexts in which it emerges. In other words, modernity is always contingent and, therefore, always distinctive.

For Giddens (1990), modernity also implies a discontinuity, a break or rupture with the past. He argues that the 'modes of life brought into being by modernity have swept us away from *all* traditional types of social order, in quite unprecedented fashion' (1990: 4, emphasis in original). To be sure, change can be profound and can lead to rupture. But as the more detailed case studies introduced below will illustrate, modernisation tends to build on what has gone before, rather than displaces it. So rather than rupture, modernisation leads to social change that reworks, overwrites, and admittedly sometimes undermines, the patterns of the past. But to understand why the experience of modernity evolves as it does in different societies it is first necessary to root an understanding of modernisation in history.

Modern lives: Thailand, Zambia and Nepal

The reconceptualisation of the nature and operation of the household and the village discussed earlier in this chapter has come about partly because scholars have looked again at established ways of thinking. However there is a second reason why this has occurred: namely, because social, cultural, economic and political change embodied in modernisation has pressured families, households and communities to adapt and operate in new ways. There is not one story to tell here, but a myriad of tales and experiences which reflect the unique ways that modernisation is inscribed on people and places, and is embraced by them. Much of the rest of this chapter will illustrate this multiplicity of experiences through looking in greater detail at case studies drawn from Thailand, Nepal and Zambia.

Than samai *cultures in Thailand*

In Thailand, the pressures on men and women, and particularly on young women, to be modern are great. At the beginning of the kingdom's economic boom, the rock band *Carabao* wrote a song – *Made in Thailand* (1985) – that lamented the new era of mass consumerism and what it was doing to Thailand and the Thais.

Made in Thailand

Made in Thailand. What the Thai do
themselves,
Sing a song with meaning, dance and
dance with style,
The *farang* [foreigners] love it, but the
Thai don't see the value.
Afraid to lose face, worried their taste
isn't modern enough.

Made in Thailand. But when you
display it in the shop
Attach a label, 'Made in Japan'.
It'll sell like hot cakes, fetch a good
price.
We can tell ourselves its foreign-made,
Modern style from the fashion
magazine.
Nobody has to cheat us. We cheat
ourselves!

(source: http://home.swipnet.se/muang_carabao/lyrics/thai/05/05-01.htm)

Mary-Beth Mills (1997, 1999) has provided one of the most detailed insights into the pressures on young, female factory workers in Thailand, many from rural areas, as they endeavour to be both modern or up-to-date (*than samai*) women and dutiful daughters. To begin with, the fact that young, unmarried women should leave home at all is surprising when viewed in historical context. Traditionally in Thailand, men were mobile while women stayed at home. Today women are at least as mobile as men, and often more so. The boundaries of accepted practice and the norms that govern what people do have shifted for a range of cultural and economic reasons. Mills sees this process putting young women under pressure as they attempt to negotiate the desire to be modern with the wish to remain dutiful. For Mills, her respondents' 'attempts to pursue *thansamay* aspirations were fundamentally at odds with their obligations as young women to rural kin. . . . every baht spent in Bangkok was, at least in theory, one less available to assist family at home' (Mills 1999: 135).

Being modern in Thailand operates at several levels. It may mean working in a clean, modern, possibly air-conditioned, environment. It means a degree of independence and autonomy. It means being able to buy the accessories of modernity including modern (Western) clothes and consumer goods from mobile phones to personal stereos. It means having a social life which includes visits to theatres and clubs. And it means having an appearance which accords, as closely as possible, to the images of feminine beauty purveyed by the media and, in particular, a pale skin (Illustration 3.4). 'Part of what draws young rural women into the city', Mills (1997: 43) writes, 'is an unspoken but powerful suggestion that they can be at once beautiful, modern and mobile'.

Bangkok can become a 'city of heaven, and a city of hell' (Mills 1997: 47) for these young female migrants from the countryside as the pressure to adopt urban consumption lifestyles collides with the obligation to be good daughters and send money home. Sometimes it is only through engaging with work beyond the village and the rice field that a daughter, in modern Thailand, can show her commitment to the family. At the same time though, the process of leaving home as part of a loosely structured household 'strategy' may begin the loosening of the natal bonds that link a young woman to her home. Esara describes the experience of Goy, a factory worker in Bangkok and a migrant from provincial Thailand:

> Each time Goy sojourned in Bangkok, the spatial distance from parental authority and the lack of family supervision further intensified the autonomy and economic independence by which she made her own decisions regarding where she would work and live and how she would manage her own income.
>
> (Esara 2004: 2000)

Illustration 3.4 Advertising on Bangkok's mass transit Skytrain. Pale skin, Western fashions and a consumer lifestyle have become objects of desire in Thailand

While the focus here, and in much of the literature, has been on the migrant experience there are also, of course, implications for those left behind. Indeed, many rural villages in Thailand appear to be 'bereft of the vitality of youth' such has been the scale of out-migration (see Chapter 7).

The simplest way to envisage what is happening in Thailand is a cascade of interlinked social and economic changes which, at times, takes on a cyclical form as migration (for example) leads to social and cultural change which, in turn, creates a context for yet further migration:

- Resource pressures in the countryside and industrialisation create the economic context in which young women take up work outside farming and beyond the local area.
- Education and changing mores oil the process by lowering the social barriers to female mobility.
- A continued commitment to the natal village and household serve to maintain the link between the migrant and her home and family even when absences are long.
- Remittances provide these women with a surprising degree of authority and decision-making power given their age and marital status.
- The experience of living away from home gives these young women the confidence and ability to challenge traditional gender norms about the roles of daughters.
- Having been away, returning home is rarely easy.
- For those left behind, the absence of young women may require men to take on new tasks.

It is important to see this sequence of causes and effects adapting, or advancing, over time as cumulative changes alter its operation. So, while migration in Thailand may have been a phenomenon of every decade since the 1970s, the causes, outcomes, contexts and compositions of those migration flows have changed in the intervening years. The process of modernisation, and migration itself, change the social and economic contexts within which it occurs.

These tangible, material responses of young women and men in Thailand are partly moulded and driven by growing resource pressures in the countryside, by widening opportunities in the non-farm sector, and by government policies. However in the background is a profound change in how people in Thailand think about their lives and their futures and the aspirations that inform the decisions they take. *Than samai* ideology in Thailand – the desire to be 'up-to-date' – is propelled, therefore, not just by the exigencies of economic change but by messages, sometimes communicated *sotto voce*, by the media and through education:

> I want to have an office job so that I don't have to be out in the sun throughout the day. A salary of 8,000 baht should be enough for me.
>
> (18-year-old high school student from the poor north-eastern region)
>
> The happiest moment of my life was when my child was accepted to work in a company in Pathum Thani province. My child has got to work in an air-conditioned office, not in a rice field like me.
>
> (45-year-old man from a remote area of the north-eastern region)
>
> A happy man is a wealthy man. We know how happy a man is by counting his material goods such as car, money, gold, and jewelry. These rich people don't have to work hard.
>
> (59-year-old widow from the north-eastern region)
>
> (Source: Darunee Jongudomkarn and Camfield 2005)

These quotations come from a quality of life (QoL) study undertaken in late 2004. It is striking how far material goods are markers of success in Thailand and, moreover, have become symbols not only of success but also of happiness – 'A happy man is a wealthy man'. The King of Thailand may be promoting a 'New Theory' of development in which he suggests that returning to a self-sufficient, farm-based economy might offer a better way forward than fast track industrialisation (Bangkok Post 1998), but his people seem intent on building glossy, modern futures.[8]

Zambian modernities

Unlike much work on Thailand, Ferguson (1999) avoids the traditional/modern binary in his monograph on Zambia and instead writes of 'localist styles' and 'cosmopolitan styles'. He also explicitly argues against placing these two styles within any historical, evolutionary sequence. This is largely, one feels, because the experience of Zambia over recent decades, as explained above, has been one of interrupted trends and reversals of fortunes.[9] A further feature of his 'styles' is that they are worn like clothes and, like fashions, can be discarded. But he also recognises that at some point the styles he writes of begin to 'stick'. He quotes Paul Mukande, a cosmopolitan mineworker who, approaching retirement was beginning to consider his imminent village return where, he thought, he would become an ordinary villager. Ferguson asks Paul Mukande, 'You wouldn't want to live like an ordinary villager?' Ferguson writes: 'After a long, searching look into my eyes, [Paul Mukande] replied: "Would *you* like to do that?"' (1999: 229, emphasis in original).

As in Thailand, Zambia has seen the flow of rural people with farm-based livelihoods to urban settings where they make their living in various non-farm occupations. Unlike Thailand, however, the incomes generated by migrants are rarely large enough to generate a significant surplus that can be channelled back to rural homes. A second difference with the Thai experience is that migrants are more likely to be male than female. Indeed the income-earning opportunities for women would seem to be markedly fewer and there are still considerable restrictions on female mobility (see Crehan 1997: 173–174). The engagement with modernity in Zambia may have brought new desires to the village but, for women at least, it would not seem to have brought markedly greater opportunities to meet those desires.

A theme which resonates with studies from many other places is that migrants in Zambia find that their engagement with urban (modern) life changes their goals, aspirations, intentions and expectations. While initially they may have harboured the notion of returning to their village after accumulating some money, as they become embedded in urban life, so this alters (Cliggett 2003). They marry, find a job and adopt the cosmopolitan lifestyles that create a gradually widening fissure, which is far more than spatial, between themselves and their villages of origin. The discardable clothes that Ferguson writes of become, in Cliggett's account, a skin with a good deal more fixity. There may also be a desire to prevent visits by and demands from rural kin becoming too onerous and draining on what little resources urban migrants are able to accumulate.

This is not to say that contact comes to a complete end – which might seem to be the logical outcome of the process. Cliggett notes that it is important that a woman returns to her husband's village to give birth to her child. It means that the very large costs of delivering a baby in the city are forgone and it also serves to reconnect the urban couple with their country kin. Migrants, therefore, find it useful to maintain links with home and do so through gift remitting – usually very small scale in terms of value and often irregular. At one level gift remitting would seem to be a legacy and a hangover from the operation of traditional family relations and would seem to make little sense in a world where money is scarce and demands are great. However it can also be seen to be sustained by some very modern concerns: 'Through exchange of material goods, however small the gifts may be, migrants express sentiments and loyalty to their rural origins, which in turn offer a hedge against the unpredictability of the contemporary socio-economic landscape' (Cliggett 2003: 549). Gift remitting has become akin to an insurance policy should any migrant find, for whatever reason, they need to return to their village of origin. Given the uncertainties and unpredictabilities of modern African life, this is only sensible. The value of the gifts may be very small indeed, even insignificant in material terms, but their symbolic function is important. Gift-remitting can therefore be seen as both old and new.

Modern medicine and dhamis *in Nepal*

Like all poorer countries, the government of Nepal, with the help of development agencies and NGOs, has been intent on introducing 'modern' medicine to the country. At the same time, particularly in rural areas, Nepal has a rich tradition of healing through *dhami-jhankris* or shamans. Stacy Pigg has written on this encounter between the modern, global world of medicine and the traditional, local world of the shaman (Pigg 1992, 1995, 1996). Rural Nepalis experience modernity in all its guises and manifestations through a development ideology which tells them that they are not modern. 'The salience of development in Nepali national society cannot be over-emphasized', Pigg writes, 'the idea of development grips the social imagination at the same time its institutional forms are shaping the society itself' (1996: 172).

Pigg tells the story of health professionals who visit a village and use the act of shamanic healing as a tool to advance modern medicine and make it more acceptable and palatable to local people. A series of dualities underpin the story: between the health 'professionals' and local people; between modern medicine and *dhamis*; and between knowledge of science and belief in *dhamis*. At first sight it might seem that the health professionals have used their guile and wit to create a victory for science and modernity over ignorance and superstition. Pigg, however, sees in this encounter something much more ambivalent:

> [The health professionals'] strategy for teaching villagers about the value of 'modern medicine' involves getting villagers to transfer their (supposed) awe towards shamans to medicine instead. In the process, medicine is presented to villagers as simply another form of magical power. . . . It is important to keep in mind that however much these representations draw on the imaginative tropes of the modernization narrative, they are enacted, as social practices, in a context that can never possibly fit perfectly within modern categories.
>
> (Pigg 1996: 178)

From a distance, the story of the engagement of Nepali villages with modernity in the guise of modern medicine would seem to be a relatively simple one of knowledge encountering ignorance (or knowledge encountering belief), of science encountering superstition, of the global encountering the local, and of modernity encountering tradition. Up close, though, it is not so clear cut and it is up close, of course, that modernity is encountered as an everyday event. From afar, rural Nepal is dotted with generic villages inhabited by generic villagers who 'believe' in shamans. While rural villages (and villagers) differ from urban Kathmandu, 'it is also true that all villages differ from Kathmandu differently' (Pigg 1992: 504). Villagers in Nepal are, generally, much more credulous that science and development usually assume and permit. Villagers live in spaces that are already 'shot through with modern narratives' (Pigg 1996: 180). They know that their 'belief' in shamans is seen as backward and they are ready and prepared for the criticism. Western medicine does not arrive unannounced and unexpected; villagers in rural Nepal know what to expect and are happy and willing to cross the line. But because their world is the world of the village, rural Nepalis must continue to acknowledge the role of shamans even when they are modern believers rather than incredulous believers. The result is not medical pluralism where the two systems are harmoniously merged, but continued separation. As Pigg says, 'It makes sense to everyone for a British doctor to come to Bhojpur to treat people's tuberculosis, but they find it a bit absurd to imagine a shaman from Bhojpur going to England' (Pigg 1995: 31).

What Pigg shows and argues, and there are clear links here with the discussion above of Thailand and Zambia, is that modernisation is not being assimilated by the people of Nepal but refashioned. Furthermore, this refashioning does not simply end with the observation that modernity takes on a particular form in rural Nepal (and everywhere else). It is also necessary to see modernity being the means by which people engage with each other. The health professionals and local villagers make connections through their varied experiences of modernity. Modernity becomes the lingua franca of connection.

Traditional, modern, hybrid: changing lives in the Global South

In their different ways, the case studies above drawn from Thailand, Zambia and Nepal all highlight the way in which new consumption practices have infiltrated people's lives and minds. Consumerism, driven by globalisation/modernisation, has become a normative goal, even for those situated at the edges of global capitalism. But while at one level we can safely say that the pressures to become modern have infiltrated virtually every crack, corner and fissure of the world, and people everywhere have become avid consumers, it is also important to acknowledge that they consume *differently*. Take just one example: Osella and Osella's (1999) article on consumption and mobility in Kerala, India. For middle-class families in Kerala, while some consumption practices are directed at the transient and the short term, there is a concentration of investment in longer term projects which may, indeed, outlive the investor. These projects include land, housing and children's education. Poor labourers on the other hand direct their consumption practices unremittingly at the short term and the transient, and particularly in the sphere of fashion – clothes, hair styles and general appearance. The middle classes view such choices and actions with some derision: 'The labouring poor, whose consumption remains severely limited, are mocked or pitied as remaining in a sort of permanent adolescence and caricatured as "like children", unable to attain mature household status' (Osella and Osella 1999: 1020). Not only do consumption practices vary between classes (and generations) in Kerala, but also a distinction is drawn between things which are consumed internally and those that are for external show. Rice, where at all possible, is local. Fashion is non-local. So while it has become commonplace to write in general terms of the spread of consumerism, a detailed examination of what, how and why people consume reveals a more complicated story.

There is always a temptation to describes 'types' and then to ascribe people or groups to these types. This creates a second temptation: to map out the ways that people change from one type to another. And finally, this then raises a third temptation: to see people moving inevitably and sequentially from one type to the next as if there is an underlying and driving telos. 'Types' (modern/traditional, civilised/primitive, etc.) are always simplifications (and generalisations) of reality. Perhaps more importantly it is also not the case that people move – as if they were moving house – from one type to another. It would be more accurate to say, and to take the metaphor further, that they are continually redecorating or extending their homes, sometimes in ways that have little permanence (painting the walls) and sometimes in ways that have more fixity (adding an extension). Moreover, in a housing estate most houses may appear essentially the same from a distance but in their details, and in particular in their interiors, they are very different indeed. When, then, scholars write of hybrid cultures or ways of living, they are glossing over the reality that all ways of living are hybrid, and are also indeterminant and temporary. Hybrid cultures are not half-way houses on a transition to something else; nor are they an end product. In writing of hybridity most scholars are merely highlighting and emphasising what is always present and always on the point of changing to something else.

The previous paragraph, though, slams up against Ferguson's 'indubitable ethnographic fact' (Ferguson 1999: 16) that for many people there *are* styles and categories, and there is also a telos. And if events, as in Zambia, might have taken a turn for the worse (as it is viewed), then this is something which needs to be corrected. To quote Ferguson again, 'the upending of the project of modernity [in Zambia's Copperbelt] is not a playful intellectual choice but a shattering, compulsory socioeconomic event . . . from the vantage point of the Copperbelt, it is about as playful as a train wreck' (1999: 253). Scholars, then, have the challenge of acknowledging indeterminacy and contingency while also realising that this is not often the way it is viewed by the participants themselves.

There is one final paradox in the various engagements with modernity described above. Modernity becomes the means by which 'non-modern' and 'modern' people and communities communicate. It is through this communication that they come to be seen, and to see themselves, as not-yet-modern. But it is also through this process of connection that they become coated with, first, a veneer of modernity that over time seeps through to the bones.

Further reading

Guyer (1981) provides a still excellent review of household and community in Africa. Many of the themes she highlights resonate in other continents. Feminist scholars have been at the forefront of problematising the household (see Folbre 1986a, 1986b) and they have also been instrumental in pushing back the frontier of research on the life course and lifecycle. Monk and Katz's (1993) introduction to their edited book is a good place to start. There has been a torrent of publications on social capital. Instrumental in getting this publication stream flowing have been Putnam (1993) and Coleman (1988). Putzel (1997) and Fine (2002) provide hard-hitting critiques of the social capital paradigm, while Bebbington (2002, 2004b) attempts a spirited defence. For an example of social capital in practice, see Bebbington and Perreault's (1999) study of Ecuador. The downloadable DFID (1999) paper is a handy summary of the concept. Giddens (1990) provides an overview of modernity while good case studies are provided by Mills (1997, 1999) on Thailand and Ferguson (1999) and Cliggett (2003) on Zambia.

Household, village, community and the life course

Folbre, Nancy (1986a) 'Cleaning house: new perspectives on households and economic development', *Journal of Development Economics* 22(1): 5–40.

Folbre, Nancy (1986b) 'Hearts and spades: paradigms of household economics', *World Development* 14(2): 245–255.

Guyer, Jane I. (1981) 'Household and community in African studies', *African Studies Review* 24(2/3): 87–137.

Monk, Janice and Katz, Cindi (1993) 'When in the world are women?', in: Cindi Katz and Janice Monk (eds) *Full circles: geographies of women over the life course*, London: Routledge, pp. 1–26.

Social capital

Bebbington, Anthony (2002) 'Sharp knives and blunt instruments: social capital in development studies', *Antipode* 34(4): 800–803.

Bebbington, Anthony (2004b) 'Social capital and development studies 1: critique, debate, progress?', *Progress in Development Studies* 4(4): 343–349.

Bebbington, Anthony and Perreault, Thomas (1999) 'Social capital, development, and access to resources in highland Ecuador', *Economic Geography* 75(4): 395–418.

Coleman, James S. (1988) 'Social capital and the creation of human capital', *American Journal of Sociology* 94(Supplement): S95–S120.

DFID (1999) *Social capital*, Key sheets for sustainable livelihoods 3, London: Department for International Development. Downloaded from http://www.keysheets.org/red_3_Soccap_rev.pdf.

Fine, Ben (2002) 'They f**k you up those social capitalists', *Antipode* 34(4): 796–799.

Putnam, Robert, with Leonardi, Robert and Nanetti, Raffaella Y. (1993) *Making democracy work: civic traditions in modern Italy*, Princeton, NJ: Princeton University Press.

Putzel, James (1997) 'Accounting for the dark side of social capital: reading Robert Putnam on democracy', *Journal of International Development* 9(7): 939–949.

Modernisation, modernity and modernism

Giddens, Anthony (1990) *The consequences of modernity*, Stanford, CA: Stanford University Press.

Case studies in modernisation

Cliggett, Lisa (2003) 'Gift remitting and alliance building in Zambian modernity: old answers to modern problems', *American Anthropologist* 105(3): 543–552.

Ferguson, James (1999) *Expectations of modernity: myths and meanings of urban life on the Zambian copperbelt*, Berkeley, CA: University of California Press.

Mills, Mary Beth (1997) 'Contesting the margins of modernity: women, migration, and consumption in Thailand', *American Ethnologist* 24(1): 37–61.

Mills, Mary Beth (1999) *Thai women in the global labor force: consumed desires, contested selves*, New Brunswick, NJ: Rutgers University Press.

4 Making a living in the Global South

Livelihood transitions

Introduction

In Chapter 3 the traditional/modern binaries that are implicit in much work on change in the Global South were brought into question (see page 57). Individuals, households and communities do not make neat and full transformations from a former 'traditional' state to a 'modern' one. Nor is there a 'traditional' state that can be easily identified and then categorised. Change is characteristic of all societies, at all times in their histories. In every case, therefore, we need to ask the questions 'from what?' and 'to what?' In writing, however, that the traditional/modern binary fails to acknowledge the messiness of change it is also true that many people in the Global South *do* have a clear idea either that they are making the transition, or that they would like to, from a traditional to a modern existence. Scholars, for a range of reasons, may find difficulties with this vision of the relationship between the past, the present and the future but people who experience it at first hand rarely have such qualms.

With this in mind, the following discussion is concerned with mapping out, examining and explaining how societies, households and individuals experience transition, and what transition means for everyday life and living. By using the word 'transition' the discussion will therefore – and necessarily – fall into the 'trap' that the world of experience sets for the scholar, namely of assuming that people move from one state to another.

From past, to present, to future

It has become common to see the past as a faulty mirror image of the present (Table 4.1). We look to current modes of existence and the inequalities, dependencies and the degree of commercialisation and monetisation which seem to be such an intrinsic and inherent part of modern lives and regard these qualities as products of processes initiated, largely, from and over the second half of the twentieth century. So, if we see current lives and livelihoods as (and increasingly) commercially oriented, outward-looking, mobile, unequal, individualistic, dependent and competitive then the corollary must be that past lives and livelihoods in the Global South were subsistence-oriented, inward-looking, sedentary, egalitarian, corporate (community-oriented), self-reliant and tranquil.

When it comes to describing and accounting for economic change three key terms are often used interchangeably, namely: commoditisation, commercialisation and monetisation. It is important, however, to recognise the differences between them. Commoditisation refers to the production of goods and services for exchange. This may be for sale or for barter. Monetisation refers to the use of a universal exchange unit – i.e. money – whereby the relative

Table 4.1 The past and the present

Past livelihoods	*Current livelihoods*
Subsistence-oriented	Commercially-oriented
Inward-looking	Outward-looking
Sedentary	Mobile
Egalitarian	Unequal
Corporate (community-oriented)	Individualistic
Self-reliant	Dependent
Tranquil	Competitive

values of goods and services can be expressed. Commercialisation, meanwhile, means the organisation of production on a commercial, business-like basis, for profit.

A brief survey of the more historically engaged work on Latin America, Africa and Asia quickly shows, however, that ordinary people in these continents were far from living in worlds 'unto themselves' or leading existences that were divorced from wider national and even global level processes. The Thai economic historian Chatthip Nartsupha may quote in his seminal book *The Thai village economy in the past* (1999 [1984]) a villager saying 'in the beginning things had no price' (page 16) but it seems that most things had an exchange value and, furthermore, that the monetisation of the Thai economy had gone quite far (in geographical terms) and quite deep (in terms of what was monetised) many years before Thailand became 'modern'.[1] There is clearly too much variety and variation to be excessively rigid in our characterisation of the past. What we can say, however, is that societies in the past were more dynamic, differentiated and integrated than hitherto imagined but, at the same time, generally not as dynamic, differentiated and integrated as they are today. These are weasel words but the point is to emphasise that we are assessing trends and relative weightings; there is little new under the sun.

Eric Wolf's (1982) book *Europe and the people without history* represents the fullest attempt to challenge the assumption that until quite recently the world was an unconnected place. As he draws the threads together of his tour of the world in 1400, he concludes that:

> Everywhere in this world of 1400, populations existed in interconnection. . . . If there were any isolated societies these were but temporary phenomena – a group pushed to the edge of a zone of interaction and left to itself for a brief moment in time. Thus, the social scientist's model of distinct and separate systems, and of a timeless 'precontact' ethnographic present does not adequately depict the situation before European expansion; much less can it comprehend the worldwide system of links that would be created by the expansion.
>
> (Wolf 1982: 71)

Since the publication of Wolf's *tour de force*, numerous other scholars have provided case studies of particular areas and activities to lend detailed support to his assertions. McSweeney's (2004) work on the dugout canoe trade in the Mosquitia region of Central America which stretches along the Caribbean coast from eastern Honduras to Nicaragua, for example, 'undermines the persistent notion that market exchange among remote rural peoples is an unprecedented [i.e. modern] activity . . . and serves as a reminder that remote peoples have been entangled with international capital circuits for centuries' (2004: 653). She notes that even the 'famously retiring' Sumu groups have been intensely involved in market exchange networks, some transnational in scope, for at least 250 years. The same message is communicated in historical studies of the island of Borneo where there was a rich, active and diverse trade in 'jungle' produce in the pre-colonial period. Luxury items such as hornbill ivory, rare woods like camphor and gahru for oil and perfumes, medicinal products, birds' nests, cutch (a dyestuff) and more found their way from the interior forests of Borneo to the coasts and from there to the sultanates of Brunei and Sulu, to China, and beyond (Cleary 1997) (Illustration 4.1). The earliest European accounts show that in the mid-nineteenth century, before intensive European intervention, luxury jungle products were 'collected and marketed in a complex and sophisticated manner' and formed part of an international trading network (Cleary 1996: 320). It seems, moreover, that the various Dayak (tribal) groups and the semi-nomadic, hunter-gathering Punan were actively and enthusiastically engaged in the collection and sale of these products to Chinese and Malay traders (Cleary 1996: 305–306). Such collection was wide and well organised, market signals were efficiently and quickly transmitted to producing areas, and the system was financially rewarding permitting a degree of capital accumulation among indigenous groups.[2] Cleary warns against making too sharp a distinction between the pre-capitalist/indigenous and capitalist/colonial periods in Borneo's history. Pre-colonial trading patterns and the products that comprised the trade flows continued to be important in the colonial period and rather than a sharp historical break we find a more subtle melding over time.

Illustration 4.1 Betel leaves and the nuts of the areca or betel palm for sale in Thailand. Chewing betel nut has long been a tradition of the cultures of Asia and the Pacific, although the habit is dying out in many places

LIVELIHOOD (AND POVERTY) TRANSITIONS

In Chapter 2 it was noted that one of the shortcomings of livelihoods research, particularly when the word 'livelihoods' is paired with 'sustainable', has been a tendency to underplay the degree to which livelihoods adapt, progress and morph over time (see page 34). The discussion in this section focuses on what is termed livelihood 'transitions'. In order to reflect on livelihood transitions, however, it is worthwhile beginning with a consideration of poverty transitions.

Poverty transitions

Statistics on poverty often give the impression of quiescence or relatively slow change. Graphical representations such as those in Figure 4.1 may show the incidence of poverty dramatically declining (China, Malaysia), remaining stubbornly fixed (Philippines), or even rising over time (Indonesia, Zambia). Occasionally, national crises may lead to relatively short-term increases in poverty, as happened following the Asian economic crisis of 1997 (Indonesia, Malaysia). In other cases the progress of poverty may reflect longer-term national economic malaise (Zambia). What these data, and the tables and graphs that are generated from the data, do not show, however, is the turbulence that lies beneath the sometimes suspiciously tranquil surface. These disturbances relate to three key areas:

- first, the effects of short-lived events on poverty such as, for example, a crop loss arising from flooding, drought or a pest attack;
- second, the intra-personal dimensions of poverty and the variability of experience within populations, and within populations over time; and
- third, the 'bottom end churning' that occurs close to the poverty line as sometimes large numbers of households oscillate between the categories 'poor' and 'non-poor'.

These concerns focus, in large part, on the poverty trajectories of individuals and individual households. This concern for individual experiences is because aggregate data on poverty tend to mask two important issues. First, and most obviously, they mask the variability in experience. As Ravallion (2001: 1812) writes, 'The churning that is found under the surface of the aggregate outcomes . . . means that there are often losers during spells of growth, even when poverty falls on average'. The second issue which the aggregate data tend to shield from view is the degree to which the poor population in any country is being continually reworked over time. So while the poor may be always with us, the individuals who make up any poor population are shifting as some individuals and households lift themselves out of poverty, while others find themselves falling beneath the poverty line. This has led to a growing interest in panel or cohort studies which track households over time.

Arising from this work, some scholars subdivide the poor into the 'persistent poor', the 'chronic poor' and the 'transient poor' (Jalan and Ravallion 2000). The persistent poor are always poor whatever the month or year. The chronic poor are usually poor but at some points in the year may find themselves above the poverty line. The transient poor, meanwhile, are those people or households who are normally non-poor but, often due to seasonal fluctuations in output or income, may find themselves dipping beneath the poverty line at some periods in the year. This final point is exemplified in the village of Khaliajuri in rural Bangladesh (Netrokona district) where the daily wage rate declines from 100–140 taka a day during the peak rice harvesting months of March and April when there is buoyant demand

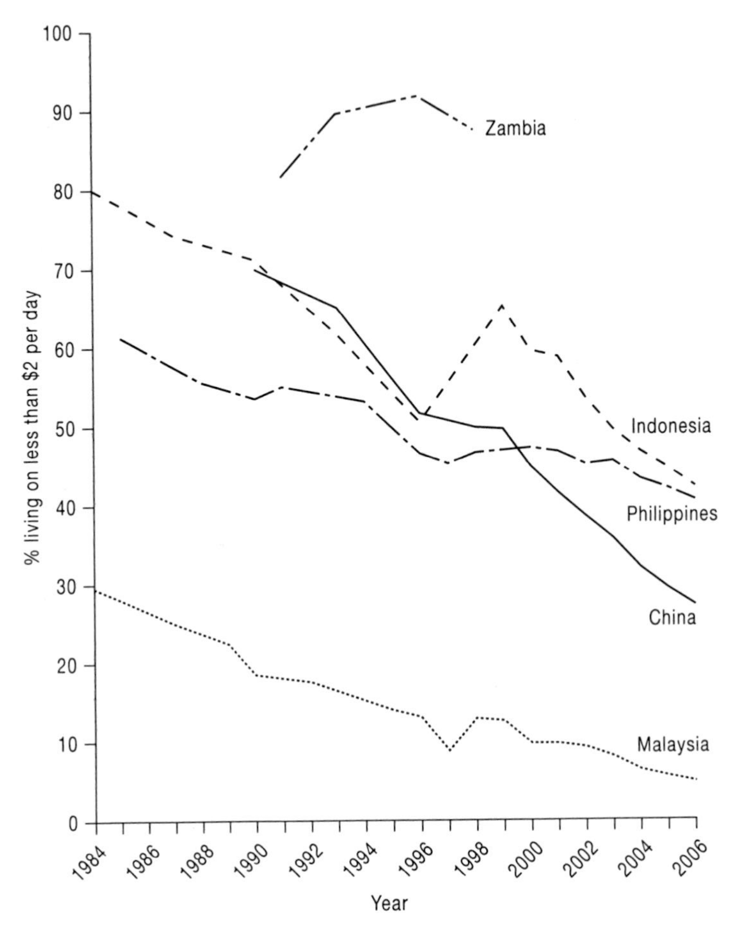

Figure 4.1 Poverty trajectories: percentage of the population living on less than $2 a day in selected countries of Asia and Africa

for labour, to 40–60 taka in the slack season, to as little as 15–20 taka in some other areas where labour is abundant and work limited (Nabi et al. 1999: 37; Narayan et al. 2000: 56).[3]

It is important to note that this identification of different poverty transitions is not just an alternative way of categorising the poor. While the persistent poor may be broadly equated with the 'bottom poor' in Table 4.2, poverty transitions aim to elucidate the processes that make people poor (or non-poor), rather than the experience of poverty per se. The degree of short-term turbulence is clear in Figure 4.2, which shows the summary results of a panel study undertaken in six villages in semi-arid South India over nine years from 1975 to 1984 (Gaiha and Deolalikar 1993). The data show that while just one-fifth (22 per cent) of households were poor throughout this nine-year period, nearly nine-tenths (88 per cent) were poor for at least one year in the nine. Another panel study, this time from Ethiopia, found that the population at risk of being poor was 50 per cent to 75 per cent greater than the actual number of poor at any one point in time (Dercon and Krishnan 2000). What these studies and others show, in other words, is that the numbers of people who *might* be poor (i.e. the potential population of poor people) is often very considerably larger than the number who *are* poor, taking a snapshot in time.

Table 4.2 Categorising the poor: state and process

Well-being category (experience of poverty)		*Poverty transitions*	
Social poor	Poor who can turn to the community for support in times of crisis	Transient poor	Usually non-poor, occasionally poor
Helpless poor	Landless; unable to afford health care or to educate their children; cannot entertain guests	Chronic poor	Usually poor, sometimes non-poor
Bottom poor	Often households headed by elderly men or women, sometimes disabled; the bottom poor have no obvious means of support and often starve	Persistent poor	Always poor

Source: Well-being categories refer to Bangladesh and are extracted from Nabi et al. 2000: 120–121

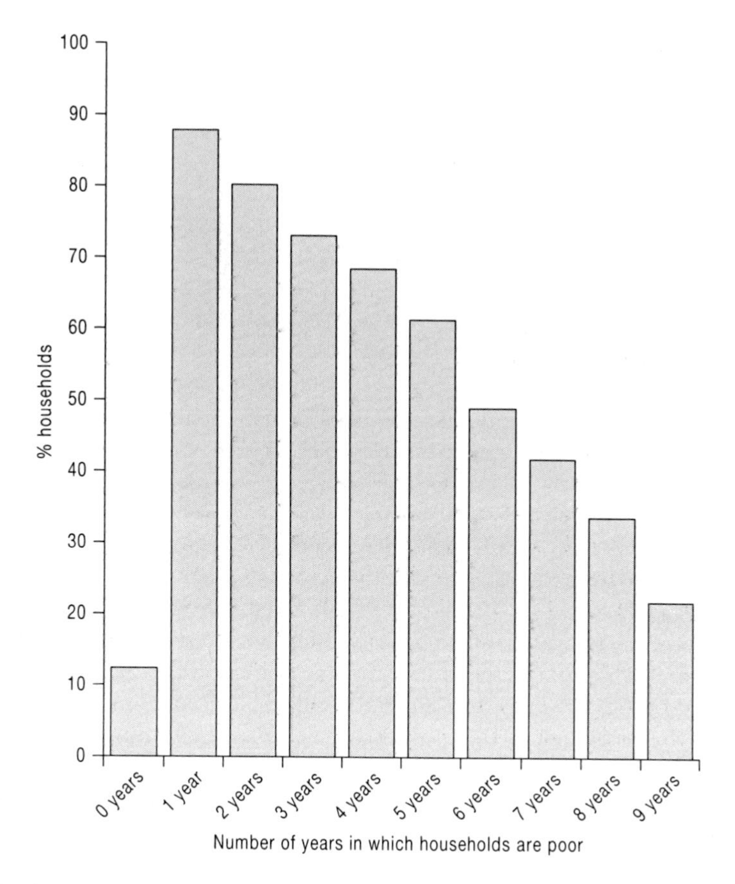

Figure 4.2 Persistent poverty estimates, rural South India (1975–1976 – 1983–1984)

The concern for 'poverty dynamics' focuses predominantly on the annual fluctuations in the status of households as they grapple with a capricious environment or dramatic fluctuations in the availability of work. Arguably of even greater interest is what happens over rather longer periods as 'economic mobility' either permits households to lift themselves permanently out of poverty or, alternatively, causes them to be consigned to the category 'poor' on a longer-term basis (Baulch and Hoddinott 2000). We can assume that the different trajectories for China and Zambia in Figure 4.1 are linked to the relative economic dynamism of these two countries. However at the village level the explanation for why some households are 'rising' and others 'falling' may be linked not to such wider economic trajectories but to the varying opportunities open to households at different stages in the life course (as noted in Chapter 3), to local resource scarcities, or to the educational attainment or skills of individuals. Once again we can make generalisations about the likely course of poverty under conditions of economic expansion or stagnation, but at a household or village level the experience may be quite at odds with such generalisations. Ravallion acknowledges this issue when he makes a plea for more micro, country-specific research so that it is possible to identify those factors explaining 'why some poor people are able to take up the opportunities afforded by an expanding economy – and so add to its expansion – while others are not' (Ravallion 2001: 1813). It is, of course, necessary and important to identify and pick out trends and patterns, but this should not go so far that it gives the impression that we can anticipate who will be poor. There are always important elements of luck, chance, fortune, contingency and serendipity involved.[4] Thus in a study of poverty dynamics in 36 villages in Gujarat during the 1990s the authors, in listing the factors that push some households into poverty (and keep others out of poverty), note particularly the role of ill-health and health care costs, and customary expenses on marriage and death feasts (Krishna et al. 2005: 1178–1179).

Participatory poverty assessments

This perceived explanatory 'thickness' gap in poverty studies has been partially filled by the increasing use of participatory poverty assessments (PPAs). Indeed, as part of the production of its 2000/2001 World Development Report *Attacking Poverty* the World Bank (2001) not only drew on PPAs already undertaken in some 50 countries but also specially commissioned another 23 PPAs which incorporated the views of 20,000 poor people (Table 4.3) (http://www1.worldbank.org/prem/poverty/voices/). The exercise went under the overarching title 'Voices of the poor', presumably to separate this study from the more usual tendency to hear the 'Voices of the experts'.

The core of the methodology which underpins participatory poverty assessments is that the poor are the true 'poverty experts' and therefore any report must begin and end with the 'experiences, priorities, reflections, and recommendations of poor children, women and men' (Narayan and Petesch 2002a: 2). The World Bank's integrative report *Voices of the poor: crying out for change* (Narayan et al. 2000) opens with a quote from a poor man in Adabiya, Ghana: 'Poverty is like heat; you can not see it; you can only feel it; so to know poverty you have to go through it' (page xvii). PPAs are also, and by definition, participatory. In other words, they draw on participatory research methods which tend to be qualitative rather than quantitative, open ended rather than tightly structured, and interactive or exploratory rather than based on a one-way flow of information.[5] A third point to make about PPAs in general and the World Bank's initiative in particular is that they emphasise the context-specific nature of poverty. There are poor people in every country of the world but their experiences of poverty, the underlying causes of poverty, and the available avenues out of poverty will be necessarily

Table 4.3 PPAs commissioned for the World Development Report 2000–2001

Europe and Central Asia	*Latin America and Caribbean*	*Africa*	*Asia*
Bosnia-Herzegovina	Argentina	*Egypt	Bangladesh
Bulgaria	Bolivia	Ethiopia	India
Kyrgyz Republic	Brazil	Ghana	Indonesia
Russia	Ecuador	Malawai	*Sri Lanka
Uzbekistan	Jamaica	Nigeria	Thailand
		Somaliland	Vietnam
		Zambia	

Note: All the national reports listed above, except for those asterisked, can be downloaded from http://www1.worldbank.org/prem/poverty/voices/reports.htm#lands

different. That said, it is also important to be aware of common threads that link the causes and experience of poverty across the globe such as social exclusion, marginalisation, alienation, isolation and lack of power.

PPAs have begun to address the important question 'Why are people poor?', which standard survey-based analyses have been unable to do, or at least not in any great detail. However, what PPAs have been less good at, it is suggested, is in viewing poverty within a temporal framework. There is certainly a concern with crises, stresses and vulnerabilities – and these have a temporal dimension in terms of how people 'cope' or 'manage' – but what is rather less evident are the adaptations, progressions and transformations that households and individuals embrace, create or manufacture as their livelihoods change over time.

Livelihood transitions

The growing concern and interest in poverty dynamics and cohort (panel) studies has been important in revealing the turbulence that lies beneath the surface of the aggregate data on poverty. What such studies, in the main, do not show is why people are poor, why they become poor and, for that matter, why and how they become rich. We can identify broad trends and trajectories, hazard an informed guess at why such trends are evident, but not – at least in detail – explain the differing experiences of households and individuals. It is here that a livelihoods perspective offers the necessary explanatory 'thickness' to complement the longitudinal dimension offered by some poverty studies. There is a direct link between poverty transitions and livelihood transition, the former being an outcome of changes in the latter.

So poverty studies and livelihood studies have rather different strengths. Poverty studies have been good at tracing change over time. Traditionally this has been undertaken at the aggregate level but rather more recently there has been a growing interest in the intra-personal dynamics of poverty. Livelihoods studies have been generally poor at providing an insight into change over time but rather better at providing the explanatory fabric explaining why households and individuals do what they do.

Insights into livelihood transitions

Livelihoods change, that much is clear. But why do they change and what implications does change have for different groups whether delineated geographically, socially, culturally or economically? It is hard to summarise the experience in the sense of coming to conclusions that have general validity for the reason, noted above, that geographical and historical context matter deeply. The discussion that follows, therefore, rather than providing a 'summary' discussion, will give a segmented presentation of case studies that illuminates the range of experiences across the Global South showing how individuals and households adapt, change, progress and transform their livelihoods over time.

The difficulty in telling these stories, however, is deciding how best to organise them, for there is no obvious 'best' organisational structure. One approach would be to write of 'rural', 'urban', and 'rural-urban' livelihoods. Another might be to structure the discussion in terms of 'traditional' and 'modern' livelihoods. Or, alternatively, to opt for a geographical division and write of 'African', 'Asian' and 'Latin American' livelihood transitions. Each has it advantages. The approach taken here, however, is to divide the discussion into livelihood transitions during 'normal' and 'abnormal' times, and during times of 'crisis'. The sampling protocol and selected case studies are set out in Table 4.4. There are evident shortcomings in this approach. To begin with, in countries experiencing general economic decline there will be upwardly mobile households with vibrant livelihoods, while the reverse will also be true. Even in villages or towns that might have been hit by a livelihoods crisis (as discussed later with reference to the 2004 Indian Ocean tsunami) there may well be households who have benefited and who emerge from the crisis stronger than before the event. This returns to the point highlighted earlier in the chapter of the need to look beyond the aggregate data and beneath the averages. A second shortcoming concerns the simple division (and therefore separation) of periods into 'normal' and 'abnormal'. It is possible to argue that all times are abnormal and that in individual communities there will be households and individuals in situations of relative stability ('normality') living next door to those who are experiencing great instability ('abnormality'). Third and finally, there is the danger of holding too much store by 'cherry-picked' examples which then become reframed as in some way representative of the situation in the Global South in general terms. The devil really is in the detail.

Notwithstanding these shortcomings, the approach taken here does serve to reinforce certain key points with which this chapter is concerned:

- There is livelihood change, sometimes quite profound, even during periods of 'normality' or apparent 'stability'. Conversely, at times of crisis livelihoods can exhibit a surprising degree of resilience.
- The propelling forces in livelihood transitions often embody a cocktail of local, national and international factors that extend from the environmental to the economic, political and social. Scholars often tend to privilege one or other set of factors according to disciplinary persuasion and the need to bring analytical bite to their work.
- Livelihood transitions operate at several levels or scales. Individuals age; households pass through a sequence of changes that have livelihood implications; economies evolve; and countries' prospects improve, falter and decline.

Livelihoods transitions in normal times: Ghana and Bangladesh

As was noted in Chapter 3, most people in the Global South continue to live in rural areas – comprising, in 2003, some 58 per cent of all developing countries' populations and 73 per cent of the population of least developed countries. Not only, however, are these figures fast declining so that by 2015 it is estimated that almost 50 per cent of the population in the Global South will be living in urban areas (http://hdr.undp.org/reports/global/2005/pdf/HDR05_HDI.pdf), but also livelihoods increasingly straddle the rural–urban divide (see

Table 4.4 Geographical and livelihood context of major case studies discussed in text

	Country	*Livelihoods context*	*Livelihood dynamic*
Livelihoods in normal times	**Bangladesh**	Rickshaw pullers, Dhaka	Health risks result in unsustainability of livelihoods in the long term
		Street children, Dhaka	Importance of social assets over material ones in determining livelihood success
	Ghana	Livelihoods in seven rural and two urban communities	Declining well-being in a context of rising wealth and declining poverty
		Livelihood change in farm households in Bawku District, north-east Ghana (1975–1989)	Livelihoods squeezed by national policies, deteriorating international trading context, and environmental change
Livelihoods in abnormal times	**Mongolia**	Livelihood adaptation during post-socialist transition	Rising poverty and vulnerability; retreat into herding and the informal sector
	Democratic Republic of the Congo	Chronic insecurity and economic crisis in Kinshasa	Surviving in the informal sector
Livelihoods in times of crisis	**Bangladesh**	Floods of 1998, northern Bangladesh	The floods cause a downshifting in well-being as the social poor become the helpless poor, and the helpless poor become the bottom poor
	Thailand	Tsunami of 2004, southern Thailand	Recovery and livelihood transformation

page 86). To put it another way, it is becoming ever harder to pigeonhole people – and particularly, households – in spatial, sectoral and livelihood terms. That said, livelihoods, especially in Africa and South Asia, continue to have an important rural focus and will do so for some years to come.[6]

The broader challenges and tensions facing rural livelihoods often relate to the following three areas (see Narayan et al. 2000: 46–51):

- *Declining access to land* – arising from demographic, environmental and political pressures reflected in the fragmentation of holdings, unfavourable land tenure politics and land degradation.

- *Declining returns to agriculture* – arising from the erosion of the profitability of farming due to the high costs of inputs, declining prices of outputs, and marketing inefficiencies.
- *Declining returns to other traditional rural activities* – arising from growing pressures on common property resources (water bodies, forests, grazing land) as they are degraded and/or sequestrated by the state or powerful individuals and groups, and increasing competition from non-local alternative or substitute products.

On paper – and simplistically – Ghana is on an upward trajectory. GDP per capita growth averaged 1.8 per cent per year between 1990 and 2003, and poverty rates have declined. However much of this expansion and improvement has been concentrated in and around Accra, the capital, and in the forest zone. In addition, a large part of the population lives very close to the poverty line so small changes in income/well-being can have very large effects in terms of the total number of poor people. In 2003, 45 per cent of the population lived on less than $1 a day, but 79 per cent survived on less than $2 a day.

The Ghana PPA was carried out under the auspices of the World Bank in seven rural and two urban communities in 1999 (Kunfaa 1999; Kunfaa and Dogbe 2000). What it shows is the degree to which, even in a country which is exhibiting only modest economic expansion, notions of 'well-being' are continually reworked and redefined. So it is not just what a livelihood is, or what it objectively delivers (in terms of income or food), and how a livelihood changes over time, but also how that level of output is regarded or perceived (Illustration 4.2). So, in all the nine PPA research sites it was reported that people considered conditions and existences in 1999 as more problematic and difficult than ten years earlier. Sometimes the reasons for this were relatively easy to pin down: environmental decline (lower rainfall, land degradation, deforestation); security issues (cattle stealing); and lack of job opportunities in rural areas. But a common theme was the growing need for 'deep pockets'. Women in the research site of Tabe Ere in Lawra district in the northern savanna zone (Upper West Region), for example, acknowledged that only in recent years had there arisen the expectation that villages should have a school, sanitation and electricity. The absence of such things is now perceived to be a problem. 'Having "deep pockets" means having clothing, modern housing, property, a car, and the means to educate children' (Kunfaa and Dogbe 2000: 42).

There have also been important changes in the contribution of women to livelihoods but this, too, is not always seen in positive terms.

> More and more women in poor communities in Ghana are directing impressive entrepreneurial energies towards feeding their families, clothing and schooling their children, and supporting themselves. . . . Unfortunately, despite their earnings, most Ghanaian women in urban and rural communities conclude that overall they are worse off than in the past because men are less responsible toward their families and are drinking more, and because it is more difficult for everyone to earn a living.
>
> (Kunfaa and Dogbe 2000: 39 and 40)

So women can – and in a sense have to – earn additional income, but this is often a source and cause of intra-household conflict.

On the basis of the Ghana PPA it seems that there are a series of changes in livelihoods – and livelihood outcomes – which appear on first impression to be contradictory (Table 4.5). On the one hand there is more income, less poverty, better human development, more responsibilities and opportunities for women, and better services. And on the other, there is more violence against women, greater turmoil in the household, greater ill-being, and a

Illustration 4.2 Fish-smoking in Ghana. Photograph: Kate Gough

Table 4.5 The up and down livelihood escalator in Ghana

Positive	*Explanatory filter*		*Negative*
	Local and personal	*Non-local and political economy*	
> Income	Escalating pattern of needs	Inequality	> Ill-being
< Poverty	Operation of the household	Arbitrary action by elites	> Precarious
> Responsibilities for women	Gender relations		> Turmoil in the household
> Opportunities	Gender relations		> Violence against women
> Human development		Market integration	> Social differentiation
> Services and amenities		Social exclusion	> Unequal access to services and amenities
> Entrepreneurship (particularly women)			

heightened feeling of precariousness. To make sense of these two apparently conflicting sets of processes and outcomes we need to view livelihoods against the non-material backdrop of changing patterns of needs and the sorts of changes in gender relations and in the operation of the household that were outlined in Chapter 3. It is through these explanatory filters that we can square, for example, declining poverty and rising incomes with greater ill-being.

What the Ghana PPA does not achieve – and this criticism could be levelled at participatory work of this nature in general – is to embed an understanding of rural livelihoods and their evolution over time in the wider national and international political economy. When a discussion group of women in Twabidi village in the Ashanti region was asked to outline the causes of poverty they provided the following list: many children, sickness, death, lack of financial assistance, poor weather, unhelpful husband, lack of help from disobedient children, laziness and an absence of monthly paid jobs (Kunfaa and Dogbe 2000: 22). What is noticeably absent here is a sense of the wider national (and international) context within which livelihoods operate.

This wider context becomes evident in Whitehead's (2002) study in which she tracks livelihood change among farm households in the Bawku district of north-east Ghana over some 15 years between 1975 and 1989 (Table 4.6). In her article she identifies three key factors driving or influencing livelihood change in the area. First of all, the deteriorating international economic context of the early 1980s; second, the effects of Ghana's structural adjustment policies which began to bite from 1984 onwards; and third, the impact of changing rainfall patterns on agricultural production. In livelihood terms it is difficult to separate out cause and effect. Environmental change was leading to a decline in grain production and a growing need to diversify into alternative activities. The decline in the domestic cocoa industry meant that male migrants were travelling to Côte d'Ivoire to find work. This, in turn, was causing women to take on greater and new responsibilities in terms of farming and crop management. Finally, households were becoming increasingly discrete economic units, less beholden to the wider compound, and to kin and community. One outcome, however, was that poorer households were finding their community safety nets increasingly frayed as the social fabric came under pressure due to intensification and commoditisation. At the same time, the support networks for the better-off were being enhanced.

Table 4.6 Tracking livelihood change in north-east Ghana

	1975	*1989*
Own account (subsistence) farming	20% of households self-sufficient in grain	5% of households self-sufficient in grain
Diversification and commercialisation of farming	Farming activity focused on production of staple crops – millet, sorghum and beans	Greater concentration on cash crops – rice, onions and groundnuts
Mobility – destination	Male migration to work on cocoa farms	Male migration to Côte d'Ivoire
Mobility – timing		Growing incidence of circular migration with men returning during the rainy season to farm
Compound/household/ family relations	Households and families embedded in large compound-organised units where kin and community relations are strong	Growing individualisation of farming and the division of households into discrete units separated from the compound
Gender relations		Women increasingly taking on 'men's work' such as cultivating cash crops

Source: information extracted from Whitehead 2002

These two sets of livelihood studies from Ghana show the importance of acknowledging the viewpoint we adopt in interpreting livelihood change. The Ghana PPA takes an unashamedly local view of threats to livelihoods and, implicitly, assumes that all necessary knowledge resides in the minds and the memories of 'the people'. It therefore highlights livelihood failures and threats which arise from personal experience – such as illness and laziness. Whitehead's (2002) study is closer to a political economy of livelihoods, emphasising the manner in which livelihood change at the local level can be seen – and explained – in terms of wider environmental, social, political and economic transformations. We can also reflect on this 'point of view' issue with regard to the apparently contradictory changes outlined in Table 4.5. The explanation offered above focused upon illuminating the changes in household and gender relations, and in community notions of well/ill-being. However if we take a political economy approach then an additional set of explanatory factors comes into view. Take just one area of investigation: the operation of land markets in Ghana.

Gough and Yankson (2000), in their study of land markets in peri-urban Accra, note how during the pre-colonial and colonial periods customary land tenure dominated, with ownership held by the collective (whether family, lineage or clan) and use based on membership of the collective (Gough and Yankson 2000: 2486). With independence, the state extended its control over land through formalising land titling and registration. This, though, led to extensive land grabbing by well-connected elites. Today, in peri-urban Ghana, land is held through a mixture of traditional (communal), capitalist (private), illegal (squatting) and public (state-owned) modes. Recognising that land markets are not operating efficiently, the trend is towards re-emphasising communal modes of ownership on the basis that they are more egalitarian, progressive and locally sensitive. This is being pushed forward through the Land Administration Project. However, while such communal systems may assist low income indigenous groups who can claim membership of the collective (but see below), migrants settling in the peri-urban zone have no such traditional claims to land. Chiefs can, though, sell leasehold interests to 'strangers', or non-indigenous migrant settlers. The money from such sales is taken as 'drink money', and not recorded.

The tensions and inequalities that arise from this system are evident. Migrant settlers are excluded from the communal land market and, if they want land, have to pay for leaseholds; women, who should have equal access to land, are excluded from decision-making while the money generated is controlled and allocated by the (male) chiefs and elders; and the young complain that their inheritance is being sold off by communal leaders (Gough and Yankson 2000: 2497). These tensions, furthermore, are becoming amplified over time. With the land squeeze in peri-urban areas and the deepening commodification of land even indigenes are often unable to acquire land unless they can pay for it.[7] The amount of money being generated from the sale of leaseholds is growing and while some chiefs and elders are very good at ploughing this money back into the community, others are not and the lack of transparency in the system inevitably generates suspicion. Customary tenure can and sometimes does work efficiently and with equity in Ghana; increasingly though, it does not (Paul Yankson and Kate Gough, personal communication). The customary land system may traditionally have functioned on the basis that 'land belongs to a vast family of which many are dead, few are living and countless numbers are still unborn' (Ollenu 1962, quoted in Gough and Yankson 2000: 2490), but this has become largely redundant, an historical artefact, in the face of contemporary change. We can, on the basis of this example from Ghana, therefore add another column to Table 4.5 ('Non-local and political economy') which illuminates the wider structural changes in economy and society and their impacts on livelihoods and well-being.

While rural livelihoods may still dominate in the Global South, many urban centres are experiencing population growth that far outstrips rates of natural increase. Rural poverty and stagnation have led to high levels of migration from the countryside so that capital cities and larger urban centres are growing, in population terms, by 3–5 per cent per year. In Bangladesh the urban population rose from 9.9 per cent of the total population in 1975, to 24.2 per cent in 2004, and is projected to reach 34.4 per cent by 2015. In the countries of East Asia relative rural stagnation is often coupled with high levels of overall economic growth, so that the figures for China's past, present and future urban population as a percentage of total population are 17.4 per cent (1975), 41.8 per cent (2004) and 49.5 per cent (2015).[8]

In Dhaka, Bangladesh, many of the poorest migrants to the city, who leave the countryside as part of a process and strategy of 'distress diversification', become rickshaw pullers (Illustration 4.3). Begum and Sen (2004, 2005) have written a series of papers examining the status and prospects of these rickshaw pullers and note that the occupation offers increased returns in the short and medium term and modest upward mobility but, ultimately, an unsustainable livelihood. This is because of the health risks of rickshaw pulling. As rickshaw pullers age, deteriorating health combined with the cumulative dangers of the work begin to erode the income advantages that it initially offers compared with other opportunities open to the poor in rural and urban areas (Figure 4.3). Among former rickshaw pullers, some 85 per cent left because they were physically unable to continue. Not only does it seem that rickshaw pulling fails to deliver a 'sustainable' livelihood but also it reinforces the intergenerational transfer of poverty because the educational profile of rickshaw pullers' children is little better than their fathers'. Moreover, when rickshaw pullers bring their families to Dhaka the enrolment rate in education actually declines.

To understand why rural people migrate to Dhaka to become rickshaw pullers it is necessary to view this urban livelihood in rural context. Over nine out of every ten rickshaw pullers in Dhaka is a migrant from the countryside. At the same time, almost two-thirds of the 500 current (402) and former (98) rickshaw pullers interviewed by Begum and Sen in 2003 owned no land; another 22 per cent owned less than 50 decimals (0.2 ha), making them functionally landless.[9] For Bangladesh as a whole, the average land holding of a farm household in 2000/2001 was 0.65 ha. For poor households the figure was 0.29 ha which, at prevailing levels of productivity, is sufficient to meet just one-third of poverty-level income (Hossain 2004: 8–9). In other words, for many tens of millions of rural Bangladeshis there is no choice but to scramble a living outside farming, which includes working in Dhaka. The rickshaw pullers interviewed by Begum and Sen were also socially deprived with most having no (58 per cent) or low (17 per cent) levels of education. It is in terms of this wider context, of general rural stagnation and the squeezing of agriculture-based livelihoods, that it is possible to understand why so many Bangladeshis take up work with apparently so little in the way of prospects. First, rickshaw pulling offers easy entry for poor rural migrants with few skills and no capital. And second, it provides an immediate and regular income of reasonable size compared with the alternatives available. But in other respects rickshaw pulling is a cul-de-sac in livelihood terms: incomes decline over time; few pullers 'escape' into more remunerative work; and the children of rickshaw pullers inherit the poverty and social deprivation of their fathers and mothers.

This is the general picture. But some pullers do manage to 'escape'. Begum and Sen (2004, 2005) argue that it is those men who leave the occupation relatively early in their pulling career (after 5–9 years in the job) and who are aged between 30 and 44 years old who exhibit the greatest likelihood of upward mobility. Another study shows just such an example of

Illustration 4.3 Rickshaw puller, Dhaka, Bangladesh. Photograph: Ann Le Mare

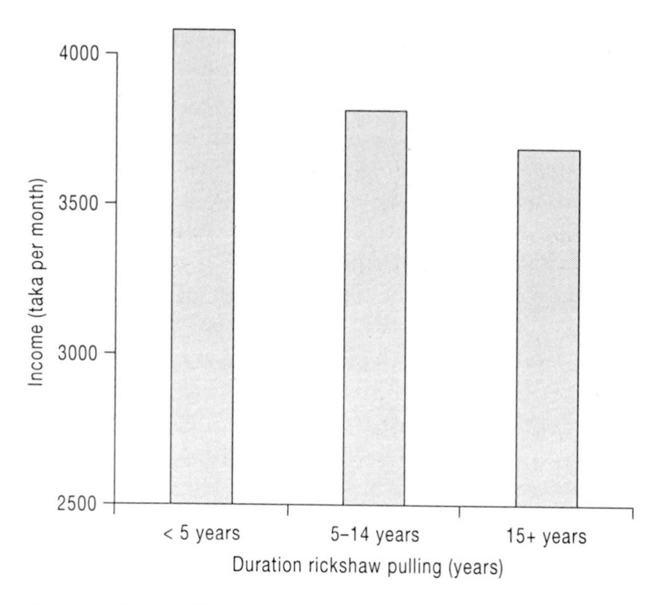

Figure 4.3 Income from rickshaw pulling and duration of work

success amidst the generality of failure (Narayan et al. 2000). Mahood Rab began pulling rickshaws at the age of 18 when he arrived in the city of Chittagong with his wife as a penniless migrant from the countryside (so far, so usual). He worked hard and managed to save sufficient money to buy his rickshaw, then buy another, until he finally owned eight rickshaws. He then borrowed money from an NGO to build rental accommodation to meet the needs of new migrants arriving in the slum where he lived. 'Mahood shared with the researchers that due to his wealth everyone knows him, and he is among those who are respected and take part in the major decisions of the neighbourhood' (Narayan et al. 2000: 52). Mahood Rab's livelihood success was reflected in a higher income and a comfortable life; it was also manifested in local respect, status and prestige.

Rickshaw pulling is a risky occupation taken up, largely, by the vulnerable who have been displaced from the land or marginalised in the countryside.[10] There are also certain groups in Dhaka – rather than occupations – who are particularly vulnerable. Street children stand out in this regard. It has been estimated that 500,000 children live on the streets in Bangladesh; 75 per cent of these, or some 375,000 children, live in Dhaka (Conticini 2005: 72).[11] Drawing on interviews with 93 street boys and girls undertaken between 2002 and 2003, Conticini argues that because living on the streets creates such a vulnerable living context, street children place greater store on non-material assets, and in particular in the social networks which deliver safety and security. Without any fixed abode, it is difficult for children to protect the material assets they might acquire. On this basis it might also be assumed that street children have little scope for improving their livelihoods and raising their well-being. It is because of this assumption that some reports refer to street children 'simply wasting their time', not acquiring skills or education, and therefore being ill-equipped to becoming 'productive members of society in the future' (CPD (Centre for Policy Dialogue) 2003: 17). Contincini (2005: 81), however, 'question[s] the inevitability of a life of destitution for these street dwellers' and sees the standard views of NGOs, policy makers and many academics failing to take on board the need to assess assets and livelihoods in terms of children's views and experiences.

A theme with general resonance across the Global South is the degree to which the rural–urban divide is becoming blurred.[12] This applies at several levels. First, it applies in terms of the sectoral delineation of activities. Industries are colonising the countryside and there is also a good deal of agriculture in urban areas (and always has been). Second, it applies in terms of livelihoods. Households and individuals are embracing both farm and non-farm activities as they seek to make a living. Occupational multiplicity or pluriactivity is becoming the norm in many areas. Third, it applies in terms of the spatial 'address' that we give to people. It is not just, in other words, that industries are colonising the countryside but rural people are oscillating with increasing frequency between rural and urban 'homes', and households are becoming spatially split. Scholars write of 'shadow' households where, from a functional point of view, there are household members who contribute significantly to the operation of the household but who are rarely physically present. What you get, in livelihood terms, is often more than what you see. And finally, it is occurring in terms of the cultural and social attributes that we assign to 'urban' as opposed to 'rural' living. Writing of rural *kampung* (villages) in Malaysia, for example, Thompson (2004: 2358) makes a case that 'Malay *kampung* are socially urban spaces, in so far as the social lives of the inhabitants conform to the characteristics of urban life'. This example is explored in greater detail in Chapter 6 (see page 133).

A key livelihood issue is whether the evolving and intensifying links between rural and urban areas and activities reflect progress or distress. The optimistic scenario is that the sort of changes sketched out in the last paragraph reflect, and are driven by, a series of positive developments: more employment opportunities, higher incomes, modernising agriculture, greater productivity, improved skills and education, better communications, and so on. On this basis, Evans outlines a virtuous cycle of farm/non-farm relations in his work on Kenya (Evans 1992; see also Evans and Ngau, 1991) where non-farm income is invested in agriculture, leading to higher farm output, rising incomes, heightened demand for local goods and services, the further development of non-farm activities, and greater non-farm employment and income generation. The pessimistic outlook, on the other hand, contends that the processes are not developmental at all but reflect rural decline and the displacement of poor people from the countryside. In this line of thinking, the rickshaw pullers mentioned earlier are not in Dhaka from choice, but because they simply have no alternative.[13]

Livelihood transitions in abnormal times: Mongolia and the Democratic Republic of the Congo

The examples of Ghana and Bangladesh reveal some of the ways in which livelihoods progress (or do not) during relatively normal periods. What, though, of livelihoods in abnormal times when events – political, economic or environmental – force change which lies outside the ambit of 'normality'?

The importance of considering livelihoods within a wider structural frame, as suggested in the context of Ghana, becomes even clearer in the case of Mongolia where the economic reforms associated with post-socialist transition have been a catalyst for a fundamental change in the character and balance of livelihoods at a national level (Table 4.7). In summary, the reforms dismantled the socialist era economic and social support networks, forcing many people to enter the informal sector or to leave urban areas and take up livestock herding on Mongolia's pastoral commons. The number of herding households more than doubled between 1990 and 1997 so that by the end of the decade, herders constituted half of the labour force. The retreat of households into herding, Mearns (2004: 115) states, 'should be seen as part of a household survival strategy, not as part of a national development strategy'. As formal sector employment and the regular wages associated with it evaporated, so households became more vulnerable, lives became more insecure, and livelihoods more complex and diverse. Studies make a loose distinction between 'coping' and 'adaptive' strategies.[14] Coping strategies include, for example, reducing consumption, the use of non-monetised inter-household transfers, recourse to the informal and illegal economies, and family 'splitting' when households divide themselves spatially (such as between rural and urban areas noted above) as a way of maximising returns and getting by.[15] Adaptive strategies include migration, livelihood diversification and livelihood switching.

Mongolia's economic reforms arose out of the liberalisation agenda which characterised the 1990s and while some of the outcomes might have been expected – for example the retrenchment in the formal sector and rising levels of inequality – the livelihood responses came as a surprise to many analysts. So too did the rise in levels of poverty, to 36 per cent of households by 1995. The National Statistical Office of Mongolia (NSOM) and the World Bank undertook a detailed study of 180 households in 2000 to identify 'triggers of impoverishment' – the shocks that push households into poverty (NSOM 2001: 21–22). What is significant here is that while the most commonly mentioned shock for poor and very poor households was loss of employment, which can be clearly linked to the reform process, the second most prevalent shock was the illness of a family member.[16] Illness is largely unpredictable and while it is possible to observe that the dismantling of social support systems makes

Table 4.7 Economic reform and livelihood transition in Mongolia, 1990–2000

Economic reforms ⇨	⇨ *Effects* ⇨	⇨ *Livelihood outcomes*
Closure of state-owned enterprises and laying-off of workers	Rising unemployment	Decline in livestock productivity due to over-stocking and rising conflicts over access to pastures
Loss of subsidies from the former Soviet Union	Urban–rural migration	Greater inequality
Retrenchment in the public sector	Growing role for the informal sector	Dramatic increase in the incidence of poverty, rising to 36.3% of the population in 1995
Collapse in state revenues	Doubling of the number of herding households between 1990 and 1998 from 17% to 35% of total households	Rising maternal mortality and declining school enrolment
	Reduced provision of health and education	Growing reliance on social networks (social capital) and inter-household transfers for the poor
		Increased engagement of children in livelihood activities, including begging and theft
		Growing diversity and complexity in livelihoods as households diversify to survive
		Family 'splitting' so that households can take advantage of opportunities in urban and rural locations

Source: information extracted from Mearns 2004

the livelihood implications of being ill more pronounced, it is not possible to identify and read off the individual households who are at risk. In terms of poverty categories there is a degree of predictability; at the level of the household and individual there is unpredictability and uncertainty. It is also worth noting that through some of the responses outlined in Table 4.7, households were able to navigate through the hard years of reform and by the end of the 1990s some of the trends had been reversed. Households began to drift back to the cities as economic growth and employment picked up; health and education profiles were rising; and levels of poverty were, at least, not getting any worse and were probably improving.

The example of Mongolia shows how policies, and their economic (and social) outcomes, can lead to profound livelihood adaptations. In the Democratic Republic of the Congo (DRC), which has experienced an even deeper economic decline than Mongolia, it is not so much government policies but endemic insecurity and the absence of policies and effective government with which the population have had to contend.

The DRC has experienced one of the world's sharpest falls in wealth and well-being with its economy contracting, year-on-year, by 3.9 per cent per annum between 1990 and 2003.[17] Over approximately the same period GDP per capita declined from US$288 in 1960 to US$100 in 2003. The only people, it seems, to have benefited from the DRC's precipitous decline are the 'entrepreneurs of insecurity', the warlords who created the conditions of chronic insecurity in the first place. As rural dwellers have fled the insecurity of the countryside, the capital Kinshasa has grown from 400,000 inhabitants at independence in 1960 to more than 7 million in 2004 (Iyenda 2005: 56). It is all the more remarkable, then, that 'in the absence of state support, with the entire economy in chaos, with most of the public and private infrastructure in ruins, and with one of the highest rates of unemployment in central Africa', people in Kinshasa continue to exist and survive (Iyenda 2005: 66–67). Iyenda (2005), in his survey of street enterprises undertaken in 2001, shows how this has been achieved through what might be termed 'bob-and-weave capitalism'. Pushed out of formal sector employment or displaced from the countryside, the inhabitants of Kinshasa have occupied, and sometimes created, the interstitial geographical and economic spaces of the city, both as producers and as consumers. The informal sector has blossomed and grown, just as the formal sector has evaporated. Barbers, cigarette sellers, money lenders, bicycle repairers, cobblers, mechanics, vegetable sellers, hawkers of fruit, cooked food sellers, and more, have filled the livelihood gap. Some vendors sit at the street side with their goods displayed on the pavement; some have stalls that they erect each day. Still others with more established businesses may rent small kiosks. Some entrepreneurs set up in alleys between established shops and there are many more 'walking' vendors who have no fixed site of sale but navigate through their chosen streets. The employment map has been reworked not through any overarching national strategy – there has not been one – but through the aggregated efforts of many thousands of individuals. Of course these efforts cannot improve people's general living conditions (which remain abject), let alone develop the country, but they do enable people to survive.

Livelihood transitions during times of crisis: Bangladesh and Thailand

Abnormal events – crises – can have the effect of up-ending the normal sequence or trajectories of change. Occasionally such crises come close to being normal, in the sense that they have a periodicity. In other cases, crisis events really do happen 'out of the blue'.

Rural households in Bangladesh have always had to grapple with a capricious environment, a product of living and making a living in a low-lying deltaic region prone to recurrent flooding (Ninno et al. 2002). It has also been suggested, however, that this traditional pattern of risk has been amplified and accentuated over recent years. A rapidly growing population has forced farmers to squeeze subsistence production from a declining area of land, while also extending agriculture onto more flood-prone land. At the same time, deforestation has, in the eyes of many, caused flooding to become more common and more severe.[18] There was a particularly catastrophic flood in the autumn of 1998 which affected 30 million people, leaving 500,000 people homeless and 1,100 dead.[19] What the flood also did was force households who might have been 'getting by' prior to the flood, into deeper poverty. Even some rich households found themselves counted among the poor. Saduadamar Hat, Aminpara and Hatya are villages in Ulipur sub-district in northern Bangladesh, one of the poorest areas of the country. All three villages suffer from flooding and were hit hard by the 1998 event. They were also included in the Bangladesh PPA undertaken under the auspices of the World Bank in 1999, just a short while after the flood (Nabi et al. 2000: 127).[20] Hatya was the most seriously affected because it is sited on the River Brahmaputra, while the other two villages are some one kilometre from the River Teesta. In Hatya, not only were crops decimated, but also roads and houses were destroyed by the floods. In Saduadamar Hat and Aminpara the damage was limited to crop land. Figure 4.4 shows that not only did the three villages have very different poverty profiles and – one suspects – poverty trajectories before the floods, but the effects of the floods were also uneven. Nonetheless the floods led to a downward shift in the livelihood profiles of the three villages. Before the floods, 13 per cent of the population were defined as 'bottom poor' or beggars; in 1999 this figure had more than doubled to 28 per cent. That said, it is important not to explain the poverty of Ulipur purely in terms of the 1998 floods. Agricultural land is being 'devoured' by the shifting Brahmaputra, households have few opportunities beyond agriculture, and land ownership in the area is also highly skewed (Nabi et al. 1999: 66). In other words, it is certainly true that the villagers of Saduadamar Hat, Aminpara and Hatya face severe environmental crises which can knock them back in livelihood terms. They also face, however, more deep-seated structural impediments to livelihood progress, which are linked to inequalities in the distribution of resources and over-dependence on agriculture.

On the morning of 26 December 2004 a powerful submarine earthquake in the Indian Ocean triggered a tsunami that led to the death of around 200,000 people in countries from Sumatra westwards to East Africa. In Thailand, 5,395 people were confirmed dead with another 2,932 people listed as missing. Unlike the 1998 floods in Bangladesh, the tsunami was something unknown to most, and unexpected by all. But like the floods in Bangladesh, it radically transformed structures and processes of social relations and economic production in affected areas. Post-tsunami reconstruction, therefore, does not mean a reconstruction and therefore a re-creation of the pre-tsunami state of affairs. For some affected individuals

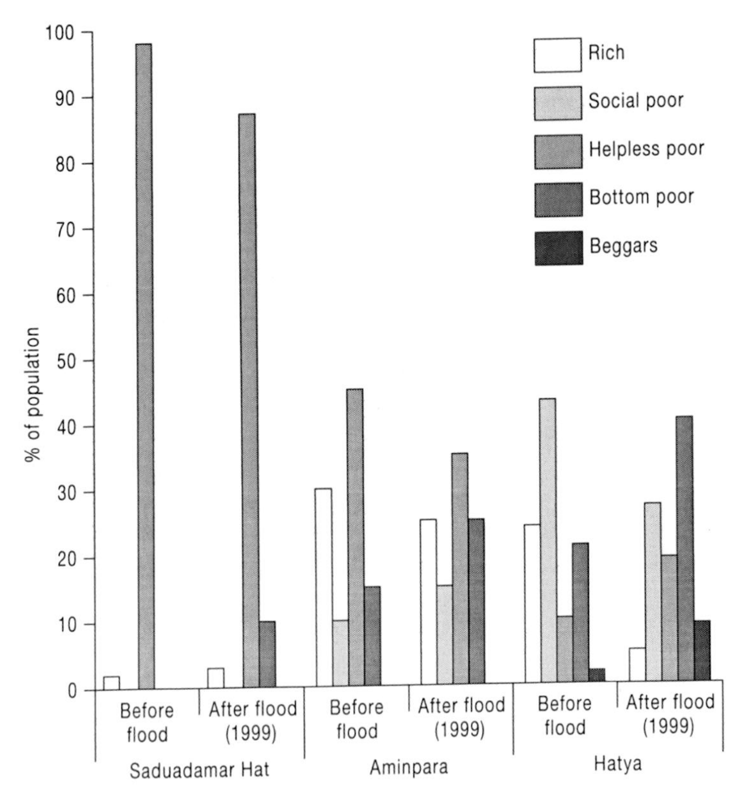

Figure 4.4 Changes in well-being in three villages in northern Bangladesh following the floods of 1998

and households the tsunami profoundly altered their livelihoods. Nor was this always for the worse; some people benefited.[21]

Before the tsunami, Gii and Ob, two Muslim sisters, ran a small pancake stall from a restaurant on the tourist island of Koh Phi Phi in the Andaman Sea where they sold pancakes, drinks and deep fried snacks. In the tsunami the sisters lost all their equipment – stove, food processor and utensils – with a total value of 25,000 baht (US$625). When we interviewed Gii and Ob in July 2005 some six months after the tsunami they had re-established themselves in a prime and larger corner plot (Illustration 4.4). While the tsunami hit the sisters with an initial loss, it also provided them with an opportunity. The monthly rental of their corner plot before the tsunami was 10,000 baht (US$250); post-tsunami they had managed to secure it for just 3,000 baht (US$75). Gii and Ob illustrate the type of reworking of social and economic space that occurs following an event of this magnitude. They have used the lower rental costs to move into a prime corner location. They have replaced their equipment and have used their luck in surviving the tsunami not just to begin again, but to begin again in such a way that they have a heightened potential for accumulation. But while Gii and Ob may – in livelihood terms – have benefited from the tsunami, the man who operated out of their new plot before the tsunami, and from where he ran a clothing store, was not so fortunate. He was badly injured, had his leg amputated, and then left the island to settle elsewhere.

Illustration 4.4 Gii and Ob serving a tourist at their new pancake house on Koh Phi Phi in southern Thailand. Koh Phi Phi was severely affected by the tsunami of 26th December 2004 and Gii and Ob's original business was destroyed

Mapping and tracing livelihood transitions

A characteristic conclusion of studies of livelihoods in the Global South is that they are complex, shifting, dynamic and contingent. It is hard to conclude otherwise when fieldwork forces scholars and others directly to engage with individuals and households, rather than just with systems, sectors and economies. It becomes abundantly clear that each household faces its own, unique contextual mosaic which combines the personal and the structural. It also becomes clear that a myriad of livelihood systems and strategies emerge from these mosaics. Finally, each livelihood mosaic is dynamic both because people respond to changing contexts, and because they wish to change their circumstances. It is tempting to see 'context' being of a higher level than 'circumstance', which is rather more personal in its location and orientation. At the margins, however, it is often hard to see where context ends and circumstance begins; they interpenetrate one another.

Economic reform programmes can have a very significant bearing on livelihood strategies – as exemplified in the case of Mongolia where thousands of households 'retreated' to rural areas to secure their subsistence.[22] At the same time, however, a blanket observation such as this plays down the variability of responses to such high level events. Different personal circumstances – related to age, gender, class, size of household, the possibility for inter-household transfers, skills, initiative, and more – will have a role to play in determining what livelihood decisions and strategies emerge from the maelstrom of reform.

It is also true that reform is not the only show in town. Local conditions and personal circumstances may push such policies into the margins of explanation. The discussion of Bangladesh, for example, demonstrated the need to view rickshaw pulling in Dhaka in terms of the rural contexts from which many rickshaw pullers emerge. The prevalence of sub-livelihood holdings, the absence of alternative non-farm opportunities in the countryside, and the attractions of rickshaw pulling itself (easy entry, quick returns) provides much of the explanatory fabric. Once again, though, the livelihood trajectories of individuals – such as Mahood Rab – often deviate to such an extent that one wonders whether there is a 'norm' which can be usefully identified.

Even in cases of generalised crisis – such as following the 2004 tsunami in the Indian Ocean or the 1998 floods in Bangladesh – it is clear that there is no generalised impact, or generalised response. Korf (2004; see also Korf and Fünfgeld 2006) examines livelihood responses in four villages affected by Sri Lanka's long-running civil war (1983–present). In each case, the war has created an unstable livelihood context, resulting in considerable vulnerability. However, this instability was experienced very differently by the villagers, and their reaction to it was also very different. So, while Sri Lanka's civil war may provide the obvious context in which to assess livelihoods, and a key locus of explanation, patterns of vulnerability and trajectories of livelihood change draw upon, and link to, much else besides.

These are the more obvious forces driving livelihood change: floods, drought, tsunamis, war, economic reform, and so forth. In such cases it is easy to see why livelihoods change, even if the question of how they change is not so self-evident. But what the chapter has also tried to show is that even in periods of apparent quiescence there is turbulence; individual livelihoods are always shifting and dynamic, never still. People change jobs, acquire new skills, bread-winners fall ill, crops are decimated by pests, children leave home, men lose their savings gambling, wives lose their husbands, women become pregnant, and so on. There are intervening occurrences which are episodic or periodic and which, collectively, may provide more explanation than the high level events. There are also wider and more persistent patterns of change to take into account – the economic dynamism of much of East Asia, for example, compared with the general stagnation or decline of much of Africa. Work patterns are changing. Women are taking on new types of work and new responsibilities. They are getting out of the house, handling money, and taking decisions – sometimes for the first time.

The last few paragraphs set out the range of factors and responses that come into play when considering livelihood transitions. It is also worth noting, to end, that scholars bring their own perspectives, preferences and predilections to their analyses. Scoones and Wolmer (2003: 11), for example, in their assessment of rural livelihoods in southern Africa, note that they are embedded in 'contested governance contexts, where politics and power relationships *must be at the centre of any analysis*' (emphasis added). There is also a fair amount else which needs to be at the centre of any analysis, such as the environment, market relations, and the operation of the household. In opting to privilege one way of looking it is inevitable that a particular gloss will be placed on the interpretation that is offered. Scholars regularly call for multi-scalar, holistic views of livelihoods but, ultimately, such calls tend to disguise analyses and interpretations which are partial in scope and biased in their concerns.[23] There are so many possible entry points when it comes to studying livelihoods that it is also true there will be an equally large number of exit points.

Further reading

The classic revisionist historical study of connection before modernity is Wolf's (1982) *Europe and the people without history*. Cleary (1997) and McSweeney (2004) provide two case studies from Borneo and Central America demonstrating the degree of economic activity and interaction that was present even in remote places and among 'primitive' peoples. See Baulch and Hoddinott (2000) for an overview of poverty transitions and Whitehead (2002) for a wide discussion of livelihood transitions drawing on work in Ghana. For various case studies of livelihood transitions, see Begum and Sen (2004, 2005) on rickshaw pullers in Bangladesh, Mearns (2004) on livelihoods and reform in Mongolia, Korf (2004) on livelihoods in a war situation in Sri Lanka, and Nabi et al. (2000) and Ninno et al. (2002) for livelihoods and natural disaster in Bangladesh. Each of these, in different ways, shows how livelihoods are being continually reworked and how they are always contingent.

Views of the past

Cleary, Mark C. (1997) 'From hornbills to oil? Patterns of indigenous and European trade in colonial Borneo', *Journal of Historical Geography* 23(1): 29–45

McSweeney, Kendra (2004) 'The dugout canoe trade in Central America's Mosquitia: approaching rural livelihoods through systems of exchange', *Annals of the Association of American Geographers* 94(3): 638–661.

Wolf, Eric (1982) *Europe and the people without history*, Berkeley, CA: University of California Press.

Poverty and livelihood transitions

Baulch, Bob and Hoddinott, John (2000) 'Economic mobility and poverty dynamics in developing countries', *Journal of Development Studies* 36(6): 1–24.

Whitehead, Ann (2002) 'Tracking livelihood change: theoretical, methodological and empirical perspectives from North-East Ghana', *Journal of Southern African Studies* 28(3): 575–598.

Case studies in livelihood transitions

Begum, Sharifa and Sen, Binayak (2004) *Unsustainable livelihoods, health shocks and urban chronic poverty: rickshaw pullers as a case study*, Chronic Poverty Research Centre Working Paper 46, Bangladesh Institute of Development Studies, Dhaka (November). Downloaded from http://www.chronicpoverty.org/pdfs/46%20Begum_Sen.pdf.

Begum, Sharifa and Sen, Binayak (2005) 'Pulling rickshaws in the city of Dhaka: a way out of poverty?', *Environment and Urbanization* 17(2): 11–25.

Korf, Benedikt (2004) 'War, livelihoods and vulnerability in Sri Lanka', *Development and Change* 35(2): 275–295.

Mearns, Robin (2004) 'Sustaining livelihoods on Mongolia's pastoral commons: insights from a participatory poverty assessment', *Development and Change* 35(1): 107–139.

Nabi, Rashed un, Datta, Dipankar, and Chakrabarty, Subrata (2000) 'Bangladesh: waves of disaster', in: Deepa Narayan and Patti Petesch (eds) *Voices of the poor: from many lands*, New York: Oxford University Press, pp. 113–145. Downloadable from http://www1.worldbank.org/prem/poverty/voices/reports.htm.

Ninno, Carlo de, Dorosh, Paul A. and Islam, Nurul (2002) 'Reducing vulnerability to natural disasters: lessons from the 1998 floods in Bangladesh', *IDS Bulletin* 33(4): 98–107.

5 Living with modernity

Introduction

Chapter 4 investigated how livelihoods change over time. This chapter explores livelihood 'states' in the modern world with a particular focus on how people embrace and engage with what might loosely be termed 'modern livelihoods'. The words 'modern' and 'livelihoods' have to be corralled within quotation marks when brought together in this way for reasons that should already have become clear. To start with, it is often hard to know where and when the modern begins and the pre-modern or traditional ends. The two have a trying tendency to merge in surprising ways. The past contains the seeds of modernity and the modern shows tendrils of connection with the pre-modern or traditional.

Though the division may often be unhelpful, it is also inevitable and insidious. So far as livelihoods are concerned, individuals and households rarely make clear and complete transitions from one way of living to another, in the process jettisoning one (traditional) livelihood in favour of another (modern) livelihood, but meld the two. Such interlocking livelihoods link, for example, the subsistence production of a staple, food crop with work in a factory, or adapted handicraft production with the cultivation of export crops. Furthermore, it is not just what people do, but how they do it and the socio-cultural context within which work – be it 'modern' or 'traditional' – occurs. Thus, and this is explored later in the chapter, factory working conditions may reproduce (and factory managers may use) traditional social structures. At the same time, the contract farming of poultry, for example, may be very modern in terms of management and the production regime it employs.

The core of the chapter consists of an examination of people's engagement with modernity from the standpoint of two areas of work. To begin with, there is an exploration of how rural lives have become touched by commodity production and, more particularly, how rural producers and workers have become implicated in wider networks of exchange that link them with corporations and consumers, often in the Global North. This is followed by a discussion of factory work, taking a perspective from the factory floor. To provide a wider context to these two thematic examples, they are preceded by a consideration of the creation of individual modernities drawing, particularly, on the experience of China.

Individual modernities

'Individual' modernity refers to the way in which contact with, participation in, and the social experience of modernity inculcates in individuals a set of attitudes and values which can be counted as modern. Social psychologists Alex Inkeles and David Smith (1974) in their ground-breaking book *Becoming modern,* based on research in six developing countries (Argentina, Bangladesh (at the time of the research, East Pakistan), Chile, India, Israel

Table 5.1 Inkeles and Smith's (1974) four characteristics of individual modernity

Characteristics of modernity	*Indicators*
Informed and participating citizen	• Identifies with the region and the nation state • Takes an interest in and keeps informed about public affairs, national and international • Joins organisations • Participates in the political process
Sense of personal efficacy	• Believes in the autonomy and efficacy of the individual and their ability to make a difference • Rejects passivity and fatalism • Works individually or in concert with others to effect change • Believes in the possibility for personal improvement
Independent and autonomous of traditional influences	• Follows public officials rather than traditional leaders • Autonomous in terms of their personal affairs, from job to marriage partner selection
Open to new ideas and experiences	• Interested in technical and scientific innovation and advances • Willing and enthusiastic about meeting strangers • Open to female empowerment

Source: extracted and adapted from Inkeles and Smith 1974: 290–291

and Nigeria), set out to identify those factors or forces which are important determinants in individual modernity. The independent variables were grouped under ten headings: personal and family characteristics; origin; residence; socio-economic level; education; intelligence and skills; occupational characteristics; factory characteristics; information-media exposure; and work-behaviour (Inkeles and Smith 1974: 49). On the basis of their research, a modern person is said to have four main characteristics, and these are outlined in Table 5.1. These characteristics, they admit, are often just alternative ways of measuring the same thing (1974: 302). However education emerged as 'unmistakably the most powerful force . . . shaping [a person's] modernity score' (1974: 304). 'In large-scale complex societies', they contend, 'no attribute of the person predicts his attitudes, values and behavior more consistently and more powerfully than the amount of schooling he has received' (1974: 133). Occupation and exposure to mass media came out as important but of second rank in terms of their importance for individual modernity. Surprisingly, city/urban living appeared to play little role, as did religion and ethnic origin. Another important conclusion of their work is that people can, and do, change in quite fundamental ways even after they reach adulthood.

Chinese modernities and social change

China has experienced the most rapid, sustained period of economic growth of any country in history. GDP per capita grew by 8.2 per cent per annum over the 28 years between 1975 and 2003. By 2040 it has been projected that China's GDP, at market exchange rates, will be larger than that of the United States (The Economist 2006b: 11). Nowhere, it could be argued, has modern living made such deep inroads over such a short period of time.[1]

The two life experiences identified as most important in Inkeles and Smith's (1974) original work (education and the organisational context of work) have, more recently, been used to investigate individual modernity in China (Inkeles et al. 1997). The fieldwork was carried out in 1990 in the area of Tianjin (a major city and industrial centre 120 km from Beijing) among a sample of farmers, industrial workers (rural and urban) and individual small-scale traders and entrepreneurs. By 1990, in China, economic reforms in the countryside were far advanced and internal migration substantial. The results of this survey were surprising to the extent that they showed no relationship between educational level and individual modernity which, as noted above, was the clearest and strongest influence in all the countries included in Inkeles and Smith's original work. The authors put this down to the enormous disruption in China's education system during the Cultural Revolution (1966–1976) and the continued arbitrary assignment of high school leavers to unskilled jobs even into the 1980s. In China, they conclude, there is only a 'tenuous' link between educational achievement and occupational attainment, with jobs being 'mediated by municipal, district or neighbourhood labour bureaus which took little account of school achievements' (Inkeles et al. 1997: 56). A second difference between China and the six countries in the earlier study is that those working on family farms were, surprisingly, more modern (according to the criteria of individual modernity noted above) than urban, industrial workers. This the authors put down to the geographical and sectoral progress of the economic reforms in China which began in rural areas and in farming (with the production responsibility system introduced in 1978). Market-oriented work environments, they argue, even if they are rural and agricultural, will be more effective at instilling modern values than state-run industrial enterprises in urban environments. So it is not just what people do, but the exchange context in which they do it that also matters. Urban workers in state-owned industrial enterprises in 1990 China showed characteristics of passivity and dependence, and a lack of initiative. Those in market-oriented agricultural enterprises (i.e. family farms), by contrast, showed qualities of self-reliance, initiative, adaptability and personal responsibility (Inkeles et al. 1997: 58).

Education, the mass media and modernity

It has long been recognised that education, particularly of girls, has an impact on development (Bowden 2002: 405). What is explored less often is the link between education and modernity. For Sherman (1990), writing of Batak parents living in the vicinity of Lake Toba in North Sumatra (Indonesia), 'having children in school is perhaps the most promising means, for themselves or their descendants, to tap or connect to the power and wealth that vitalize the world beyond the border of the lake' (Sherman 1990: 50).[2] Children not only acquire through education the skills to access opportunities in the modern economy – skills which they can, in a very real sense, 'sell' – but also are given the driving desire to do so. For rural dwellers, farming has become unattractive not just because it often delivers lower incomes but also because it has become a low status occupation to be, if at all possible, avoided. Parents and their children have been inculcated with the belief that education, above all else, should be sought, valued and pursued. In Malaysia, a person without education is *bodoh* – stupid, uneducated and ignorant. Kelly conducted a survey of 123 school leavers in Tanza High School in the province of Cavite south of Manila in the Philippines and found not a single pupil who expressed an interest in working in farming (Kelly 1999b: 70). 'For the vast majority, working in a factory or office or going overseas are their preferred options. . . . This applies even to those whose parents are farmers' (Kelly 1999a: 297). I found the same in a survey of lower secondary school leavers in a village school in the northern Thai district of Sanpathong where just one pupil expressed a wish to farm (Illustration 5.1).

Education is not alone in instilling in people the attractions of a modern life and modern work. The media and consumer cultures more generally also make a significant contribution.

Illustration 5.1 A local, village school in Sanpathong district, northern Thailand

The pressure and desire to be modern, then, is transmitted by a trinity of influences: by state policies that promote the fact and the fiction of modernisation and development; by education and the content of the syllabuses that inform the educational experience; and by the words and images transmitted by radio, television, film and the printed media. Taken together, this trinity forms a highly powerful force for change.

LEADING MODERN LIVES: FARM AND FACTORY WORK IN THE SHADOW OF GLOBALISATION

To repeat, individual modernities are tied up with where people work and, perhaps more importantly, how people work. Global economic change over the post-1945 period, and increasingly so as the decades have passed, has incorporated hundreds of millions of people in the Global South into national and international networks of labour, production and exchange. For many scholars, officials and activists, *the* central question is whether this progressive and deepening integration and incorporation has delivered cumulative gains for participants in the process. For some, fuelled by the logic of competitive austerity (see page 113), what we see is 'immiserising growth' where greater economic activity produces lower returns and, therefore, reduced real wages (Kaplinsky 2000: 120).[3] On the other side of the debate are the defenders of globalisation who argue that the challenge facing the world is not to control or stymie globalisation but to extend and deepen it. In his widely read book *Why globalization works*, Martin Wolf (2004: 4) writes that 'the failure of our world is not that there is too much globalization, but that there is too little'. What the discussion material below will suggest is that, to put it (over-)simply, there is good globalisation and bad globalisation and any blanket assertion one way or the other will tend to hide this important – indeed, critical – fact. We should be concentrating our attention on identifying, interpreting and understanding those instances when global economic integration leads to rising incomes and improving living standards and those cases where it does the reverse rather than cherry-picking the evidence to argue one way or the other.

Internationalising agriculture, farming and farm work

Rural producers have been engaged in market and non-market exchange systems for many years. That said, the emergence, growth, expansion and deepening of global agro-food systems has changed, in a quite fundamental manner, what production for the market means, how it is undertaken, and what relationships are forged in the process. Just some of these changes are set out in Table 5.2. It should be noted that they encompass the consumption as well as the production of food in the Global South (and, for that matter, the Global

Table 5.2 Rural revolutions: changes in rural lives and production systems

Nature of change	*Case studies/examples*
Production	
Cultivation of new, non-traditional or exotic crops	• Cultivation of non-traditional export crops (snow peas, broccoli, cauliflower, French beans, mini-aubergines and various berries) in the central Guatemalan highlands (Hamilton and Fischer 2003). • Cultivation of non-traditional crops including potatoes, strawberries and sweet peas in Burkina Faso (Freidberg 2003) and flowers in Kenya (Hale and Opondo 2005)
New divisions of labour	• Feminisation of farming in Syria (Abdelali-Martini et al. 2003) • Incorporation of rural farmers into agro-food production networks in Mexico (Sanderson 1986)
New labour relations	• Expansion and casualisation of wage labouring in South Africa (Kritzinger et al. 2004).
Marketing and retailing	
Quality control and standardisation of production	• Stringent production standards for potato growers in Ecuador (see Reardon et al. 2005: 50 and Table 5.4)
Contract farming	• Contract farming in the Punjab, India and agribusiness normalisation (Singh 2002)
Integration into global circuits of capital	• Casualisation of work on fruit and cut flower farms in South Africa and Kenya (Kritzinger et al. 2004; Hale and Opondo 2005)
New retail networks	• Spread and growing role of supermarkets (Reardon et al. 2005)
Changing role and influence of the state	• Declining autonomy of the state in Mexican agriculture (Barkin 1990, 2002)
Consumption	
Dietary changes	• Increase in local consumption of milk and milk products. In China, with rising incomes and growing exposure to Western habits, per capita consumption rose threefold between 1977 and 2000 (Webber and Wang 2004: 724) • Consumption of exotic vegetables in Burkina Faso (Freidberg 2003) • Undermining of household (and national) food security

Table 5.3 The contract farming balance sheet

Positive effects of contract farming	*Negative effects of contract farming*
Farmers gain access to new technology and inputs	Agribusinesses gain indirect control over the land
Farmers can source additional capital/credit to develop their production	Agribusinesses shift some of the risks of production to farmers
Contracts with guaranteed or minimum prices shift some of the risk from risk averse farmers to agribusiness	Contract farming leads to mono-cropping and over-specialisation
Contract farming helps to develop new skills	Contracts are biased in favour of agribusiness firms, reflecting the greater power of such firms
Contract farming provides more reliable and higher incomes, permitting investment and planning	Services provided by agribusinesses are over-priced
Contract farming allows small farmers to remain on the land and in agrarian pursuits	Product prices are low
Farmers are partially compensated for crop failures	Payment is delayed
Payment is swifter than through traditional marketing networks	Contract farming favours larger/wealthier farmers over smaller/poorer farmers and causes inequalities to widen through (for example) minimum acreage stipulations
Prices are higher than with traditional marketing networks	Contract farming undermines food production and the cultivation of staple crops
	Contract farming leads to self-exploitation as farmers work longer in the fields
	Women carry an ever-larger 'double burden' of productive and reproductive work
	Women are squeezed out of traditional production activities as men take control of increasingly commercially oriented systems
	Contract farming undermines traditional communal systems of reciprocity and exchange
	Farmers have little bargaining power
	Contractual arrangements and production systems encourage over-cultivation and environmental degradation
	Knock-on effects for non-growers can be severe (e.g. if cash crops displace staple crop production and lead to higher prices of food commodities)
	Over time, farmers are progressively squeezed by agribusiness
	Contracts demand stringent quality standards that are hard for smaller farmers to meet

Sources: information extracted from Watts 1992; Little and Watts 1994; Singh 2002; Hamilton and Fischer 2003 from extension services to subsidised credit (Singh 2002). Over time, however, as the firm becomes established, prices are lowered and standards are raised as agribusinesses 'begin to rationalize grower numbers by retaining only those growers who can supply better quality produce at lower prices, and squeeze growers as they become dependent on contract farming operations' (Singh 2002: 1632).

Kaqchikel Maya farmers in the central Guatemalan highlands have been cultivating non-traditional agricultural export (NTAX) crops since the mid-1970s (Hamilton and Fischer 2003). This began with snow peas, broccoli and cauliflower, later branching into French beans, mini-aubergines and various berries. Based on survey and ethnographic work undertaken between 1998 and 2001 in two villages in the municipalities of Tecpán and Santa Apolonia, Hamilton and Fischer (2003) challenge some of the more critical perspectives on such smallholder production and argue that 'Maya farmers largely view NTAX production as a positive step towards economic advancement and as an opportunity to use their lands and labor in ways that preserve community and reinforce key elements of their cultural heritage' (2003: 82). These new crops and new markets have, somewhat surprisingly, permitted Maya smallholders to preserve their agrarian lifestyles because they represent one of the few ways in the vicinity for smallholders to meet their escalating cash needs. The authors found that there was little evidence of women being excluded from such production and, indeed, women were fully involved in land-use decisions in households producing NTAX crops.[5] Overall, the changes wrought following the introduction of NTAX crops 'were perceived [by smallholders] as overwhelmingly positive' (Hamilton and Fischer 2003: 97; see also Figure 5.1). In short, smallholders could meet the needs of increasingly demanding international buyers, large operators had not squeezed smaller producers out of the

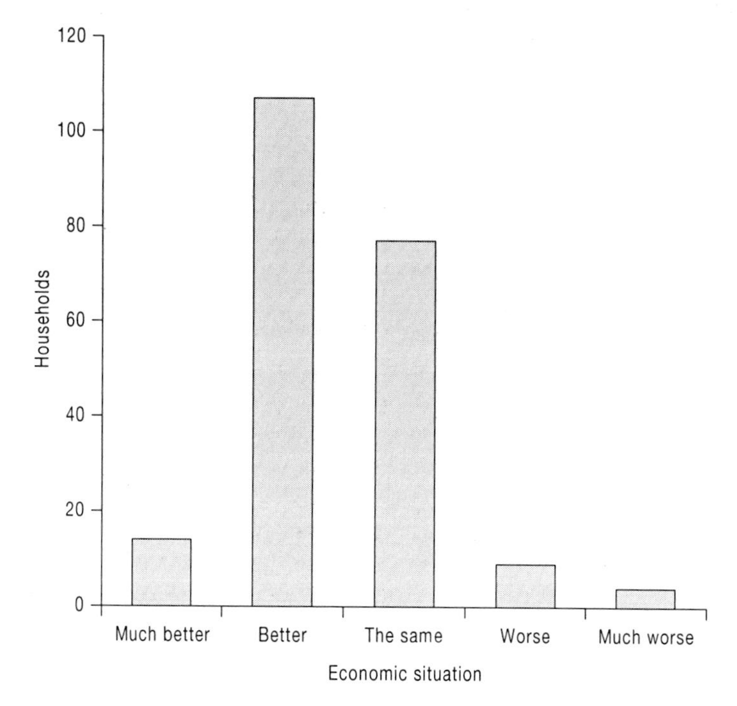

Figure 5.1 Perception of household trajectory following introduction of NTAX crops into two villages in Kaqchikel region, Guatemala (2001)

market, households were able to maintain cultivation of their traditional *milpa* crops for home consumption while engaging in NTAX production, and small-scale producers were able to maintain control of their land.

While it seems that smallholders can and do produce for the international market and are also involved in national supermarket supply chains, this does not mean that involvement in such international value chains is scale neutral. Rather, the likelihood is that degree of involvement is linked to other types of assets, beyond land. In particular, it seems that human and organisational capital are important in facilitating entry into such supply chains. This need not operate only at the individual level; an effectively functioning farmers' association, for example, may facilitate engagement and permit the bridging of the gap between agri-business demands and smallholders' skills.

Implicit in much of the above is the assumption that the production of non-traditional crops is geared to export. Increasingly, however, agribusinesses are forming contractual relationships with growers to supply the domestic market. The entry and growing role of supermarkets in the Global South, for example (see page 106), is both squeezing out traditional retailers and forging new production relationships. The ability of supermarkets to cut out intermediate links in the marketing chain brings the production demands of the supermarket direct to the farm gate. Just as contract farming makes new requirements and demands of farmers, so too do these supermarkets.[6] This is clear, for example, in the case of potato cultivation in Ecuador where supermarkets are highly exacting of farmers (Table 5.4).

Table 5.4 Exacting standards: comparing traditional and supermarket standards for Ecuadorian potatoes

Traditional market standards	*Supermarket standards*
Certain varieties	Certain varieties
Maximum level of physical damage	Maximum level of physical damage
Minimum size	Certain size
Certain colour	Certain colour
	Certain form
	Certain level of cleanliness
	Level of food safety
	Certain odour limit
	Certain maturity
	Temperature maintenance
	Specific packaging
	Certain volume
	Timing and place restrictions
	Specific payment period

Source: research by Zamora 2004, quoted in Reardon et al. 2005: 50

Working for export in the Rural South

The discussion so far has focused on producers – farmers – in the Global South. What, though, of agricultural workers? It is possible that global agro-industrialisation may benefit producers, small and large alike, while harming the workers who are employed by these producers. Hale and Opondo (2005: 304) write of the need to 'humanise the supply chain' by recognising how market pressures and demands are transmitted down the value chain from consumers in the Global North to affect the living conditions of workers in the Global South. The emphasis here is not on the product, but on the people in the chain.

Until the 1990s most UK supermarkets purchased cut flowers through Dutch auction houses. Today, they increasingly source flowers directly from producers, shortening the marketing chain and bringing consumers 'closer' to the product they are purchasing, in this instance cut flowers. Perhaps in recognition of the ever more intimate connection between supermarkets in the North and producers in the South, most UK supermarkets subscribe to codes of conduct that cover both technical (chemical usage and environmental management, for example) and social (workers' welfare) issues.

The production of cut flowers for export in Kenya dates from the mid-1960s and expanded dramatically from the early 1990s. The industry currently employs between 40,000 and 50,000 workers and is the country's second largest agricultural foreign exchange earner. Most workers are young women. To humanise the value chain, Hale and Opondo (2005) undertook 100 semi-structured interviews and 13 focus group discussions with cut flower workers in Kenya. They found that insecurity and irregularity of work were major issues for the (largely) young and female workers they interviewed (2005: 309–313). Other concerns included supervisor abuse, sexual harassment, low wages, lack of non-wage benefits (such as maternity leave), compulsory overtime, and exposure to chemicals. While Hale and Opondo acknowledge and applaud the supermarkets' role in recognising workers' rights they argue that supermarkets 'have not confronted the much bigger issue of the relationship between their own pricing and production schedules and the abuses of workers' rights in flower supply chains' (2005: 317). We can trace some of the central worries of workers interviewed by Hale and Opondo back through the value chain to the demands of consumers in the UK and the way in which these have been interpreted and transmitted by the supermarkets.

Another study of workers, this time in the fruit export industry in South Africa, also identifies a dominant trend towards the casualisation of employment (Kritzinger et al. 2004). But in South Africa the reforms since the Apartheid era have been at least as important in reworking paternalistic employment relations as have the effects of global integration. During the Apartheid period most workers were housed on farms and permanently employed. 'Modernisation' in South Africa has meant the dismantling of these structures and a shift from employment of permanent farm-based labour to flexibly and seasonally employed labour based off-farm. Rather ironically, workers in Kritzinger's study are said to look back with a sense of loss on the security of the (Apartheid) past when workers enjoyed various non-wage benefits including housing, medical care, a crèche and so forth. One worker, Rosie, said to the survey team: 'I wish I were back on the farm with all the benefits and facilities' (Kritzinger et al. 2004: 27).

From production to consumption

Most studies of contract farming and the integration of producers in the Global South into international commodity networks stress the economic, social and livelihood implications of the changes wrought. There are also, though, changes in consumption patterns and preferences in producing countries as new crops make their presence felt in terms of what people eat. We should look, therefore, not only at how work has changed but also how this has fed through into changing habits in the countries concerned.

In the countries of French West Africa, settlers and colonial officials planted gardens almost from the moment they arrived, determined to preserve the taste as well as the memory of home. The Bobo-Dioulasso region was identified by the French as the *panier* or food basket of the colony of Burkina Faso and as early as 1903 non-traditional crops such as potatoes, strawberries and sweet peas were being cultivated around Bobo-Dioulasso town (Freidberg 2003). The area under such crops expanded and, as it did so, local people developed a taste for European crops and foods. This included not only the local (non-French) elite who took to drinking *café au lait* and eating baguettes from street stalls but also ordinary villagers who would consume potatoes and other European garden crops that were surplus to colonial needs. Commercial market gardening expanded through to the 1980s, drawing farmers into the production of non-traditional crops and local people into their consumption. Freidberg is at pains to emphasise that the changes in the Burkina Faso periphery should not be equated with Westernisation. Local people continue to eat traditional staples and their traditional dishes have not been supplanted but adapted to incorporate the new ingredients available.

Much more recently there has occurred a change not just in what people eat in the Global South, and how they eat it, but also in how they shop and where they buy their food. This is most dramatically reflected in the spread of supermarkets. Reardon et al. (2005) identify three waves of diffusion beginning in the early 1990s (Table 5.5). The two key points to note about Table 5.5 are, first, the speed with which supermarkets have entered the food retail market in the Global South and, second, the range and number of countries that are included in this retail revolution. Much of Africa may still be out of the supermarket loop, but even here – in countries like Uganda, Tanzania, Mozambique and Angola – South African supermarket chains have begun to make significant investments. Furthermore, in South Africa and Kenya, supermarkets 'have spread beyond the middle class into the food markets of the urban working poor' (Reardon et al. 2005: 47). While in Africa supermarkets may have only relatively recently begun to make inroads into the traditional retail sector, in Latin America

Table 5.5 Three waves of supermarket diffusion in the Global South

Wave	*Periodicity*	*Countries or regions*	*Penetration (share of food retail market)*
First wave	Early to mid 1990s	South Africa, Costa Rica, Chile, South Korea, Philippines and Thailand	10–20% in 1990 to 50–60% by 2000
Second wave	Mid to late 1990s	South-East Asia, Central America, Mexico	5–10% in 1990 to 30–50% by 2000
Third wave	Late 1990s to early 2000s	Nicaragua, Peru, Bolivia, Vietnam, China, India, Kenya, Zambia and Zimbabwe	10–20% by 2003

Source: information extracted from Reardon et al. 2005: 45–47 and much of Asia, supermarkets have become the preferred places to shop for those living in larger urban centres.

Changes in lifestyles across the Global South have played an important role in creating the consumer context for the rapid diffusion of supermarkets. With urbanisation and the increasing participation of women in the labour force there is often not the time to shop and prepare food in the traditional manner. In the tambon or subdistrict of Khaan Haam in the Central Plains of Thailand, for example, minimarts selling processed and prepared foods have replaced traditional markets and home cooking for many households, and 'mobile' markets bring prepared food direct to workers who are short of time but relatively rich in cash (Illustration 5.2). In 1988 when the nearby Rojana Industrial Estate was established, the villages in the area were rice-growing communities where people consumed home-cooked food utilising household-cultivated rice, fruit and vegetables, captured fish, and home-raised poultry. Now, less than two decades later, most households grow no rice, vegetables or fruit, catch no fish, and raise no livestock. There is also a supply side to the explanatory equation: the spread of supermarkets can also be linked to the liberalisation of foreign direct investment and advances in supply networks and logistics that have provided the means by which supermarket chains have entered the food retail sector in the Global South.

In summary, we can see across the Global South an array of changes in agriculture and food which extend from the crops cultivated, through to cultivation systems, contractual obligations and requirements, marketing arrangements, working practices, retail systems,

Illustration 5.2 Pre-cooked and packaged food being sold to factory workers during their midday break in northern Thailand

and consumer preferences (see Table 5.4). These do not apply everywhere, and there is considerable variability between countries and continents. Nonetheless, there is a pattern in this mosaic of change which can be summarised as embodying the following: commercialisation, commoditisation and internationalisation. We can also add a second tier of terms which are not quite so generally evident but which nonetheless have wide currency across the Global South: differentiation, accumulation, specialisation, integration, acceleration, deregulation, standardisation and politicisation.

Factory work in the shadow of globalisation

The structural changes in the economies of the countries of the Global South are reflected in linked changes in the structures of employment. Admittedly the picture is far from one-dimensional, but a feature of these shifts is an expansion of factory work, much of it geared to export and a significant proportion of this financed through foreign direct investment. As was observed at the start of this section, a central question is whether such work delivers real benefits to those employed in such factories, or whether it is a case of 'immiserising growth'. This apparently simple question is not easy to answer, even though commentators on both sides of the debate may seem to suggest that it is: 'The benefits of export-led economic growth to the mass of people in the newly industrializing economies are not a matter of conjecture' (Krugman 1997).[7] That such factory work delivers substantial increases in national income and employment is not disputed. But is such aggregate expansion also translated in to higher wages, better working conditions, and improved living standards? In addition, is it possible to come to a conclusion on this undoubtedly important issue before first setting out to understand the context in which factory work occurs? One of the reasons why the question is difficult to answer is because it needs to be contextualised and set against a set of wider issues connected with the operation of the household, cultural norms, and national trajectories of change. Nor can 'better' simply be reduced to a calculation of real (economic) returns to work; there is much to quality of life beside income (see page 36).

The shoe industry in the Sinos Valley in the Brazilian state of Rio Grande do Sol expanded rapidly from the 1970s with employment almost tripling between 1970 and 1980 to 76,000 workers, and then doubling during the 1980s. But while employment may have increased substantially over the two decades to 1990 the evidence is that real wages fell. Schmitz (1995) notes the difficulty of computing real wages when inflation is high and deflators vary but nonetheless concludes that during the period 1970–1980 wages remained at 'roughly the same level' and between 1980 and 1990 actually declined by the order of 40 per cent (Schmitz 1995: 22). On the face of it this would seem to be an example of immiserising growth, where the indefatigable logic of the global economy causes wages to be squeezed as factories in centres of export production around the world compete with each other. The outcome is 'competitive austerity' and a 'race to the bottom' in terms of wages. The market-friendly approach promoted by the World Bank and other multilateral organisations becomes, therefore, reworked as 'not-so-friendly-to-labour'. Whether things are quite as clear as this line of logic would seem to indicate is returned to at the end of this section, after a foray into the villages and factories of Indonesia.

From village to factory: modern lives in Indonesia

Rebecca Elmhirst has worked for many years in a Lampungese village in the Sumatran province of Lampung, in Indonesia (Illustration 5.3).[8] In Tiuh Baru she has witnessed the

Illustration 5.3 The Lampungese village of Tiuh Baru in southern Sumatra, Indonesia. Most young women from this staunchly Islamic village leave home and travel to Java to work in the factories around Jakarta. Photograph: Rebecca Elmhirst

emergence of factory work as a key employment activity among young, unmarried women and the evolution of that work through a period of rapid economic expansion (1994–1996), Indonesia's *krismon* or economic crisis (1997–1998), the forced resignation of President Suharto (1998), and the democratisation of the country (1998 onwards).

Young women from Tiuh Baru began to travel to Java to work in the garment and textile factories in Tangerang on Jakarta's urban fringe in the early 1990s, when the first Tiuh Baru factory migrant left the village. By 1994, around one-third of all women aged 15–20 were working in Tangerang. In 1997, just before the onset of *krismon*, the figure was close to 100 per cent, numbering more than a hundred migrants in all and with just four stay-at-homes. The young women from Tiuh Baru have become part of a globalised workforce, made all the more (apparently) homogenous by the uniforms they wear and the working environment and conditions they share. Nonetheless, they remain 'first and foremost . . . *orang Lampung* – Lampungese people' (Elmhirst 2004: 389). Indeed this identification with their place and culture of origin has, if anything, hardened in recent years because their home region is a transmigration site where families from Java have settled, heightening the sense of ethnic identity among the original inhabitants of the area.

At the outset it is necessary to emphasise the cultural context within which this remarkable growth of factory work has occurred. Lampung is a staunchly Islamic area and young unmarried Lampungese women are guarded and controlled in myriad ways. They are under the guardianship of male family members; their local mobility is closely controlled and sharply limited; and they are only permitted to work in the fields in situations of dire necessity, as to do so is regarded as unseemly and degrading. Yet within a period of less than ten years long-distance migration to work in modern factories in Java became a normal, indeed the expected, activity for young unmarried women. To understand this surprising – given the norms associated with female work and mobility in Tiuh Baru – development it is necessary to consider two separate and, seemingly, contradictory areas of explanation: first, the degree to which the village has been able to extend its control into the factory work arena; and, second, the extent to which the factory is 'out of sight' and, therefore, 'out of mind'. There is also a third issue, which concerns the changes that have been wrought to 'tradition' or *adat* (custom) by the experience of factory work.

Elmhirst (2004) outlines the way in which the presence of strong village networks linking the source community with the destination site has persuaded and reassured parents that their factory daughters are being adequately supervised. Such networks are not just, however, networks of surveillance and control; they also act as support networks and social safety nets, easing and oiling migration and the process of finding work and a place to stay. While these networks may provide some reassurance to the parents of migrant daughters, there is also a sense, and this is the second issue, that working away from the village permits a degree of latitude and freedom that would not be tolerated if the work was locally situated. On this basis, Elmhirst believes that another part of the answer to the conundrum outlined above is that 'there is no shame in what cannot be seen'. One daughter said to Elmhirst, in explanation, 'my parents do not mind what I do in Tangerang, they just care what I do in the village'.

It is also true that the experience of factory work has not merely been accommodated within existing social structures and cultural norms, but has also served to stretch them: 'I have tentatively suggested that . . . among Lampungese, spatial restriction on daughters has loosened, with *adat* reworked to accommodate work in Tangerang – at a distance from the village' (Elmhirst 1995a: 22). From the first few brave female migrants who challenged accepted norms, migrant work remarkably quickly became an established rite of passage for young women and a way in which a particular kind of Lampungese femininity could be developed (Illustration 5.4) (Elmhirst 2002: 157). A discourse justified such work in terms of a set of binaries that contrasted factory work with work in the fields and then linked them to (in part, new) notions of Lampungese femininity (Table 5.6).

In the mid-1990s when migrant work of this type began, young women were effectively 'assumed out' of the household and they were not expected to remit money to their families. Towards the end of the decade the remitting of funds had become established and normalised. It was not, however, invested in agriculture or other productive activities but used to buy gold jewellery, furniture, kitchen utensils and other goods that would help the migrant set up home as a married woman when she returned home. Remittances were, therefore, used largely for the benefit of the future household that the migrant would become part of on her return, with her assumed marriage. By the late 1990s, the experience of working in Tangerang had actually improved the marriage prospects of young women, rather than compromising them. But this underpinning logic was not to last.

The economic crisis of 1998 introduced another change to remittance behaviour and norms as migrant daughters became increasingly important as sources of household income during sharply constrained circumstances (Elmhirst 2002: 159–162). They became more closely implicated in household livelihood strategies. When women were laid off from work and returned home parents would encourage them to find new work. 'All she does is sleep and eat' was a common parental view of these returnees, a perspective that would have been unthinkable just five years earlier when Elmhirst began her work in the village in 1994.

Hancock's (2000, 2001) work on Sundanese female workers in Java similarly shows the effect of Indonesia's economic crisis on factory workers' contribution to household finances.[9]

Illustration 5.4 A factory daughter returns to her home village in Lampung, Sumatra, Indonesia. Photograph: Rebecca Elmhirst

Table 5.6 Factory work, field work and femininity

Field work	*Factory work*	*Lampungese femininity*
Dirty	Clean	Feminine
Heavy	Light	Lightly muscled, smooth skin
Outside, in the sun	Inside	Pale skin
Traditional	Modern	Educated
Manual	Skilful	Technologically proficient
Local	Urban	Clever, aware, sophisticated, cosmopolitan
Production of traditional agricultural commodities	Production of international branded goods	Modern, up-to-date

Source: information extracted from Elmhirst 2002

The proportion who allocated *all* their income to support the household rose from 11 per cent before the crisis to 30 per cent in 1999/2000 and in 16 per cent of cases they had become de facto their family's main income earner (Figure 5.2). Working in a modern factory environment had, itself, raised the status of these young women; and the crisis served to increase it further still. Young women were becoming more centrally involved in decision-making and in developing household livelihood strategies. Somewhat ironically, the economic crisis provided an opportunity for young women in Banjaran, the site of Hancock's fieldwork, to challenge traditional patriarchal systems.

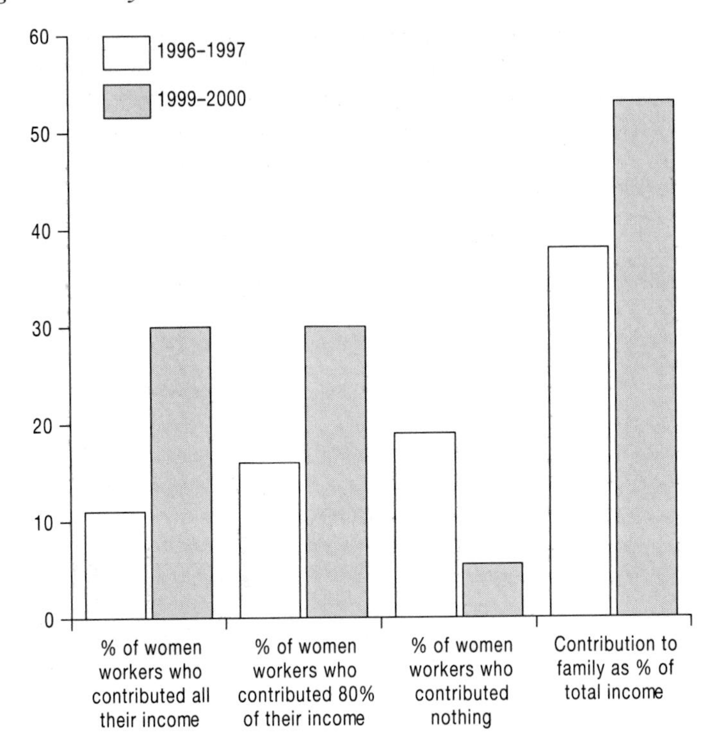

Figure 5.2 The changing contribution of women factory workers to household finances in Banjaran, Java, Indonesia (1996–1997 and 1999–2000)

At one level it is clear that to understand factory work in Indonesia it is necessary to embed any interpretation simultaneously in an understanding of rural life and living, and vice versa. Family/household dynamics and links are critical in determining the pattern of engagement with such work and, equally importantly why, in the Indonesian context, the natal bond has remained strong. Women are not just willing workers and producers; they are also reproducers and therefore the maintenance of community and links with the natal village are as important as the desire and need to earn money (Saptari and Elmhirst 2004: 38). The Lampungese migrant daughters will, almost certainly, return to Tiuh Baru, marry and settle down. Elmhirst (2004) takes this further to argue that the geographical cleavage between the source village and the place of work is, to some extent, an illusion. The village and its norms, rules and support networks are present in Tangerang, exerting their influence on activities and actions. In this way we can qualify the significance of 'leaving' and 'returning home' in so far as important elements of home are to be found in the destination sites and, increasingly, the experience of factory work is having an effect on ideas and perceptions at home.

The danger here, though, is that if this is pushed too far then migration and factory work become packaged as something entirely 'new', representing a break with the past in economic, employment and, therefore, social and cultural terms. Yet women in many areas of Indonesia, Java particularly, have always had a considerable degree of autonomy. Furthermore there was an important non-farm manufacturing sector before these factories colonised the rice fields and peri-urban spaces of Java. As Jennifer and Paul Alexander (2004: 231) highlight, we need to be careful not to assume that the legions of young, uniformed women disgorging from factories at the end of the working day represent a break with the past. It is equally possible to see factory work as the continuation and extension of an established pattern of work (see Box 5.1). There is a second way in which we can question the degree to which the Tangerang experience (and others like it) represents a break with the past and that is in terms of the distinction between *gender roles* and *gender relations*. For some scholars the changes we see in Tangerang are emblematic of the former, as young women take on new activities. However these new gender roles have not fundamentally changed prevailing gender relations (see Chant 1996; Silvey 2001: 43; see also page 132). For Silvey, 'rather than understanding gendered labor dynamics in deterministic relation to patriarchal structures and economic restructuring, analyses should conceptualize gender geographies as mutually constitutive of the specific historical forms of these very processes' (Silvey 2003: 150).

BOX 5.1 Replicating tradition in a modern setting: the case of China

China has seen an unprecedented expansion of factory work and the incorporation of young women and men, many from rural areas, into modern, industrial complexes. As the discussion of Inkeles et al.'s (1997) work made clear, how people work is an important factor determining individual modernity. It also needs to be recognised that we see in these modern workplaces not only the emergence of modern identities and ways and modes of living but also the reproduction or insertion of traditional/old identities and norms into an apparently modern context. Indeed, in some cases, traditions are harnessed to achieve very modern outcomes.

When rural migrants enter the urban labour market – which often occurs, in itself, because of long-standing socio-cultural traditions of how a 'good' daughter should behave – they often encounter a patriarchal, in some senses Confucianist, working environment which replicates conditions in their rural places of origin (Fan 2003). Wright (2003), for example, in her work in a factory in Dongguan on the Pearl River Delta in Guangdong province explains how the tight control and surveillance of female workers is linked to managers' belief that they are acting as 'factory fathers' who have a responsibility to protect their factory daughters on behalf of the young female workers' parents: '[periodic] pregnancy tests, pelvic exams, segregation and mobility policies subjected women workers to the utmost discipline and surveillance in the name of parental duty' (Wright 2003: 298). But this high level of control was also justified on the basis that it would benefit quality control by helping to control their sexual thoughts and, therefore, keep their minds on the job.

There are aspects of the Lampung/Tangerang experience which are unusual even in an Indonesian context, for example the distinctive cultural milieu, and the village networks of control which extend their tendrils to the factory context. Nonetheless there are wider implications associated with the discussion above.

Modern, factory lives: exploitative or developmental?

The opening paragraphs of this section on factory work noted the experiences of workers in the shoe factories of the Sinos Valley of Brazil's Rio Grande do Sol, and the stagnation and then decline in real wages during the 1980s. How can this apparently persuasive picture of immiserising growth be challenged?

The first point to (re)emphasise is that there are varied experiences of factory work. This variability operates at a range of levels: between countries, between industries, between companies, between domestically versus foreign invested operations, and even between workers in the same factory. But putting aside this important point that the experience of factory work varies, there is also a set of issues which lies beyond real wage rates but is nonetheless highly important in providing a rounded picture of factory lives.

As the discussion of factory workers in Indonesia shows, such work can empower young women, giving them greater autonomy and status, and a heightened sense of pride, value and self-worth. The same point has been noted in Bangladesh where employment in export-oriented garment factories has led to considerable personal change. One NGO activist remarked that the 'garment sector [in Bangladesh] has brought a silent revolution for women in our society' (Oxfam 2004: 18). These are no mean achievements particularly when they are set against a patriarchal household context.

It is also necessary to view wage rates in factories in the context of wages in other sectors of the economy and in other industries. The evidence is that there is a wage premium in foreign invested factories which also extends to domestically invested factories when wages are compared against, for example, those offered in domestic work or in agriculture (Bhagwati 2004: 172; Wolf 2004: 236). It has also been noted that while individual wages may be falling, these are not always translated into lower *household* incomes or, therefore, into a rise in poverty. This was true, for instance, of Schmitz's (1995) work in the Sinos Valley where he found that because the labour force participation rate was rising (i.e. more young people were being drawn into the labour force) there were more income earners in each household. This meant that household incomes were rising even while individual incomes were falling.[10]

It has also been noted that the attraction of factory work is not restricted to the sometimes meagre wages that are earned. Writing of Ida who, like Elmhirst's Lampungese, works in one of the garment factories of Tangerang, Warouw says:

> For Ida, self-presentation is not merely an uncritical mimicry of images and beauty rituals she learns from pop magazines or television *sinetron* increasingly accessible by the rural population. Her appearance represents her quest for modernity, a quality not available in the rural setting where she grew up.
>
> (Warouw 2003)

Factory work not only delivers the means to become modern – through regular wages – but also has come to symbolise modernity from the uniforms that such workers wear, to their pale faces and skill in working with modern technology. The state's investment in education and its efforts to instil in the Indonesian population the importance of *maju* – or modernity – find their inevitable outcome in young women like Ida who abandon one type of life for another.

It is tempting to see female factory work, whether it is undertaken in Brazilian shoe factories, Indonesian garment factories, or Malaysian semi-conductor factories as part of a household 'strategy'. It is also the case, however, that these young female workers may struggle against the expectations that are placed upon them (see Wolf 2000). They may also use their experiences of working away from home, the confidence that comes from it, and the power that money bestows, to press the moral envelope of accepted norms and practice.

Making lives modern: immiserising modernities?

It is difficult to draw the threads of this chapter together because, ultimately, the experience of the Global South is one of mixed fortunes, incomplete transitions, contingent experiences and uncertain outcomes. We are caught in a tyranny of mixed experiences where against one narrative of increased vulnerability and precariousness and of declining fortunes we can match another of apparently sustained improvements in well-being. It may be tempting, looking across the case studies from Asia, Africa and Latin America, to draw a trans-continental distinction in their general tenor. The Asian experience is one where global economic integration fuelled and driven by the neo-liberal policies since the late 1970s has created new opportunities, raised incomes, reduced poverty and strengthened livelihoods. The African and Latin American experience, by contrast (and in general), has been one where livelihoods have become increasingly precarious and where poverty appears to be as entrenched as ever. But this is unsatisfactory in the way that it labels assorted countries and places their variable populations into grand continental categories. Rather than attempting unsatisfactorily to summarise these mixed messages, I want to conclude by reflecting on some work by Eric Thompson which relates to a single village in the Malaysian state of Perak.

Thompson (2002, 2003, 2004) argues that rural *kampung* (villages) in Malaysia have become de facto urban spaces in terms of production, consumption and social interaction (Thompson 2004: 2357; see also Kahn 2001). The third column of Table 5.7 sets out the elements that Thompson, drawing on research in 'his' kampung of Sungai Siputeh in Perak, believes are

Table 5.7 Designating rural-ness (tradition) and urban-ness (modernity) in Malaysia

	Rural-ness	*Urban-ness*
Production and livelihoods	Subsistence-oriented	Negligible subsistence activities
	Agriculture-focused	Predominantly non-agricultural
	Singular, relatively unstratified economy	Diverse, stratified economy
	'Simple' lives with few consumer goods	Cash intense and consumer good rich
Social and physical interactions	Informal, personal interaction	Formal, role-focused interaction
	Relatively immobile	Relatively mobile
	A world unto itself	A place where integration and interaction is intense
	Gotong-royong (mutual cooperation) is the basis for village activities	Contract-based work and ersatz, manufactured *gotong-royong* spirit
	Attachment to the village	Alienation from the village
Identity	*Orang kampung* (people of the village)	*Melayu Baru* (New Malay)
	Backward-looking, ignorant, uneducated, lacking motivation	Forward-looking, aware of the world, sophisticated, disciplined, educated

Source: information extracted from Thompson 2003, 2004

characteristic of a state of urbanism, as opposed to that of rurality. Putting aside the fact that Sungai Siputeh is situated in the countryside – and in that sense is *in* the rural – in terms of its economic and social functioning, and increasingly in terms of its structure, it is urban. By urban – although he does not explicitly state this – he also implies modern. The New Malay or Melayu Baru who has emerged in Malaysia under the political imperative of the ruling party and with the economic stimulus of *pembangunan* (development) is 'educated, knowledgeable, sophisticated, reliable, disciplined, trustworthy, and skilled' (Thompson 2003: 428). To stretch this case study further still, we see in Sungai Siputeh the fuzziness and contingency of modernity. The village is a traditional rural space surrounded by farmland but one that is populated by modern 'new' Malays. Traditions of mutual help or *gotong royong* are regarded as emblematic of traditional village life yet *gotong royong* activities have become commercialised to such an extent that one return migrant commented that 'the apartment block [in Kuala Lumpur, Malaysia's capital] is more like a *kampung* than the *kampung*' (Thompson 2004: 2372). Young women from Sungai Siputeh not only commute to work in modern factories but also return each evening to live with – and under – the moral surveillance of 'the village'. It is because of these sorts of complexities and contingencies that it becomes hard to be anything but reluctant when it comes to describing what 'modern' lives are like and whether they are 'better' than those they have supplanted.

Further reading

There is a large literature on the effects of global market integration on farmers, farming and farm workers. Hale and Opondo (2005) and Kritzinger et al. (2004) focus on farm workers on export farms in Kenya and South Africa. Hamilton and Fischer's (2003) study of Guatemala is a generally positive view of the livelihood effects of global market integration on small scale producers and this can be usefully contrasted with Barkin's (2002) much more critical and downbeat work on Mexico. The edited volume by Little and Watts (1994) is one of the best collections of studies of contract farming; Singh (2002) looks at contract farming specifically in the Punjab. For studies of factory work, Hancock (2000, 2001), Warouw (2003) and Elmhirst (2002) all provide different takes on female workers employed in the factories of Jakarta's peri-urban zone. These are interesting to read together because of the very different approaches they adopt and viewpoints they take. For a generally balanced overview of the effects of global integration on ordinary people, download the Oxfam (2004) report *Trading away our rights*.

Global agro-food systems and the internationalisation of agriculture

Barkin, David (2002) 'The reconstruction of a modern Mexican peasantry', *Journal of Peasant Studies* 30(1): 73–90.

Hale, Angela and Opondo, Maggie (2005) 'Humanising the cut flower chain: confronting the realities of flower production for workers in Kenya', *Antipode* 37(2): 301–322.

Hamilton, Sarah and Fischer, Edward (2003) 'Non-traditional agricultural exports in highland Guatemala: understandings of risk and perceptions of change', *Latin American Research Review* 38(3): 82–110.

Kritzinger, Andrienetta, Barrientos, Stephanie and Rossouw, Hester (2004) 'Global production and flexible employment in South African horticulture: experiences of contract workers in fruit exports', *Sociologia Ruralis* 44(1): 17–39.

Little, Peter D. and Watts, Michael J. (eds) (1994) *Living under contract: contract farming and agrarian transformation in sub-Saharan Africa*, Madison, WI: University of Wisconsin Press.

Singh, Sukhpal (2002) 'Contracting out solutions: political economy of contract farming in the Indian Punjap', *World Development* 30(9): 1621–1638.

Factory work

Elmhirst, Rebecca (2002) 'Daughters and displacement: migration dynamics in an Indonesian transmigration area', *Journal of Development Studies* 38(5): 138–166.
Fan, C. Cindy (2003) 'Rural–urban migration and gender division of labor in transitional China', *International Journal of Urban and Regional Research* 27(1): 24–47.
Hancock, Peter (2000) 'Women workers still exploited: revisiting two Nike factories in West Java after the economic crisis', *Inside Indonesia* 62 (April–June): 21–22.
Hancock, Peter (2001) 'Rural women earning income in Indonesian factories: the impact on gender relations', *Gender and Development* 9(1): 18–24.
Warouw, Nicolaas (2003) 'Keeping up appearances: manufacturing workers in Tangerang make a special effort to look good', *Inside Indonesia* 75 (July–September). Downloaded from http://www.insideindonesia.org/edit75/p25warouw.html.
Wright, Melissa W. (2003) 'Factory daughters and Chinese modernity: a case from Dongguan', *Geoforum* 34: 291–301.

Globalisation and immiseration

Oxfam (2004) *Trading away our rights: women working in global supply chains*, Oxford: Oxfam International. Downloadable from: http://www.oxfam.org.uk/what_we_do/issues/trade/trading_rights.htm.

6 Living on the move

Introduction

One of the recurring themes of this book is the heightening, widening and intensifying levels of mobility that characterise the lives of people in the Global South. This mobility does not only embrace the movement of individuals, whether for work, play or out of necessity. It also extends to commodities, products, capital, diseases, organisations, knowledge and ideas. That said, the focus here is on human mobility and, more particularly, on migration. Sheller and Urry (2006: 207) observe that 'all the world seems to be on the move' and they set out a 'new mobilities paradigm' which is said to be 'spreading into and transforming the social sciences' (2006: 208). The essence of this 'mobility turn' is set out in the first column of Table 6.1.

The increased interest in and concern for the various manifestations of mobility is to be welcomed. Perhaps at long last transport geography will no longer be the subject's least sexy sub-discipline! However, the new mobilities paradigm outlined by Sheller and Urry (2006) concerns itself with what are seen to be the emerging qualities of mobility – the airport encounter, the Internet, spreadsheet culture, virtual travel, I-pods – driven by advances in technology and new ways of living and interacting. Inevitably this means that the Global North is taken as the primary site of interest and research and the space where such new mobilities are emerging. What, though, of the Global South? Is this yet another case where the Global South has to patiently wait before the ripples of change wash on its shores?

There are at least two ways in which we can think of their being – or being a need – for a new mobilities paradigm with a Southern focus. First of all, in the sense that we need to look again at old mobilities and established assumptions about the nature of pre-modern and early modern life. For example, there is good reason to challenge the 'immobile' or 'sedenarist' peasant paradigm there is so embedded in much of the literature (see page 71). And second, for the reason that while ordinary people in the Global South may not be leading networked existences akin to the lives of residents in North America and Europe, they are far from being wholly insulated from innovations occurring elsewhere. New mobilities are also arising in the Global South. These new 'Southern mobilities' not only separate the present from the past – in the sense that we need to view them as outcomes of historical processes – but also separate the Southern experience from the Northern experience. In an attempt to set these out, the second column of Table 6.1 provides an alternative, Global South-focused new mobilities paradigm. It may not be as theoretically engaged as that set out by Sheller and Urry (2006), but it does pinpoint an array of changes and challenges which are no less important in understanding the new and emerging geographies of the Global South.

Table 6.1 Outlining new mobilities paradigms for North and South

A Northern new mobilities paradigm	*A Southern new mobilities paradigm*
Social science theory and mobility • Much social science has been a-mobile; place-based theorisations need to be replaced by theories which think beyond place • Much social science has been 'sedentarist' • Mobility needs to be viewed as 'soft' embracing people, ideas, attitudes, money and outlooks – and not just concerned with the 'hardware' of mobility • Taking a mobile viewpoint requires a reconsideration of the role of scale	**Social science theory and mobility** • Theorisation needs to maintain a concern for place but not be place-based or place-obsessed • Conceptualisations of 'tradition' and the pre-modern experience have been inherently sedentarist and it is necessary to see migration and mobility as historically rooted • Theorisations of change need to accept that gender, class, ethnic and other social relations are co-constituted across space • There is a world of flows but these are not, as Appadurai (1999: 230–231) says, 'coeval' or spatially consistent; they have 'different speeds, different axes, different points of origin and termination'; unpicking and understanding these disjunctures is central to the mobility project • Theories of migration and mobility need to take cognisance of multiple drivers and a mosaic of contexts
Products, infrastructures and mobility • Roads, cars and petrol stations • Aeroplanes, airports • Mobile phones, email, computer networks	**Forms of mobility** • The growing ease of transport is permitting an intensification of mobility; a transport revolution is underway • Established gender divisions between who is mobile and who is immobile are being eroded and/or reworked
Social science methods and mobility • What are the objects of inquiry? • The 'infrastructure' of mobility embraces the networks of communication, the physical infrastructure of movement, and the attitudes that energise and are energised by mobility • Mobility does not just connect people and places but creates its own and distinct social and economic spaces (in airports and along roadsides, for instance) • The need to 'follow' people requires that methods also become mobile, for example in the form of 'mobile' or 'itinerant' ethnography	**Social science methods and mobility** • Understanding mobility should be contextualised at the level of the household/family rather than at the level of the individual • Non-local influences and effects are making their presence felt in local spaces and systems; social relations are co-constituted across space and context • The boundary between the rural and the urban (and industry and agriculture) is becoming more porous and fuzzy in terms of the people, activities and world views that are seen to be representative of rural and urban spaces • 'Shadow' households are emerging where families are divided across space

continued

Table 6.1 continued

A Northern new mobilities paradigm	*A Southern new mobilities paradigm*
Practice, experience and mobility • Geographies of mobility are, inevitably, hybrid geographies	**Mobility and livelihoods** • Mobility is leading to the progressive delocalisation of livelihoods • Mobility and gender – women are increasingly breaking the mobility envelope • Established gender divisions of labour are being eroded and/or supplanted

Note: not all the categories above under each column map onto each other. This emphasises the different areas of concern among researchers working on the South and North.

Sources: the Northern new mobilities paradigm are extracted and developed from Sheller and Urry 2006; the Southern new mobilities paradigm are based on material discussed in the chapter.

Mapping out the mobility revolution in the Global South

Earlier chapters in this book have already challenged a number of shibboleths regarding our understanding of pre-modern, traditional lives. Chapter 3, for example, raised questions about how scholars have conceptualised the operation of the village and, more particularly, the village 'community' (see page 47). The same chapter also highlighted and questioned the ingrained assumptions that underpin the meta-narrative of modernisation and the binaries that inform it (primitive/civilised, traditional/modern) (see page 57). Chapter 4, meanwhile, challenged the view that pre-modern societies were 'worlds unto themselves', isolated from the market and wider exchange networks (see page 71). We can apply the same scepticism to the view that populations in the Global South were inherently immobile – the so-called sedentarist paradigm. De Haan (1999), in a review article, presents evidence from Asia, Africa and Europe to challenge this sendentary bias and argues that 'population movement [was] the norm rather than the exception' (1999: 7). Gardner and Osella (2004: xiii) similarly warn against the assumption that migration is an outcome of modernisation, and note the historical rootedness of the process.

While there is increasing evidence to support the view of revisionist historians that pre-colonial populations were more mobile than hitherto thought, this does not take away from the fact that across the Global South there has been a marked increase in levels, types and intensities of mobility. The first question – and it is often left either unasked or taken as self-evident – is: why are so many more people moving further and with greater frequency than ever before? As one would expect, there is no single, simple answer but embedded in this mobility revolution are a set of interlinked, sometimes self-reinforcing factors. Table 6.2 presents these under five headings: resources and environment, economic, social, political, and technological. At the most mundane level, without better roads and cheap and ubiquitous transport the mobility revolution would, for many people and places, have been impossible (Illustration 6.1).[1] But of course, people either have to *need* to move – to take advantage of those transport opportunities – or *want* to move. In some cases, local and national authorities put institutional and other barriers in the way of movement (Box 6.1 and see page 159). Rather more commonly, cultural norms would seem to restrict mobility. This is particularly true, at least in the past, of daughters/young women who found that certain assumptions about acceptable behaviour limited their opportunities to be mobile. Not only do these

Table 6.2 The 'mobility revolution' in the Global South: causal factors

Category	*Causal factor*
Resources and environment	Land squeeze in rural areas
	Decline in the productivity of farm land due to environmental degradation
	Growing inequalities in access to land driven by the wealth-concentrating effects of market-led growth
Economic	Declining terms of trade between farm and non-farm activities
	Real decline in the value of agricultural commodities
	Rising costs of farm inputs
	Rising availability of non-farm work
	Increasing needs
Social	Changing gender norms regarding the acceptability of mobility, especially for (young) women
	Gradual change in the position of farming to becoming a low status occupation
	Mobility as a route to emancipation
	Growing wants driven by consumerism
Political	Lifting or easing of barriers to movement (e.g. in China with reference to the *hukou* system (see Box 6.1) and in South Africa with the lifting of the Apartheid-era pass laws)
Technological	Better roads and faster travel
	Cheaper travel
	Improved communication between sources and destination areas in terms of work opportunities and living conditions

influences and effects vary between countries and contexts, but also they vary over time and between households and individuals in particular contexts.[2]

The issue of the tension between wider temporal changes and divergent personal experiences is reflected in Rogaly and Coppard's (2003) work in the Puruliya District on the Chottanagpur Plateau of West Bengal. In the 1960s and 1970s poor people in this region of low agricultural productivity and few employment opportunities had little choice but to migrate seasonally to find work. Migration was driven by distress. In the 1980s and early 1990s, however, as work opportunities expanded and real wages increased so migration became a means by which migrants could accumulate cash and, on their return, invest this in better housing, land, livestock, and the education of their children. As one district council chairman explained to the authors, 'they used to go to eat, now they go to earn' (Rogaly and Coppard 2003: 428). There is, therefore, a historical story we can tell about how wider economic changes in this area of West Bengal have fed into changes in the nature of mobility. It is not, of course, just these wider economic changes of which we need to be aware. Working among labour migrants in Calcutta, de Haan (2004: 202–203) notes how the emphasis on education means that life for young men is no longer so relaxed. It has become more

Illustration 6.1 The mobility revolution – a public bus in Vietnam

competitive and disciplined and evenings are spent with private tutors cramming for the future rather than living for the moment.

But while this might be the overall narrative in Puruliya District, 'by switching the scale of analysis to the individual level, it is clear that much is missed in the aggregated perspective – and much of what is missed matters' (Rogaly and Coppard 2003: 428). Thus, for Soma Mahato, a young woman of 28 who already had twelve years of mobility under her belt, migration was an emancipatory process which permitted her to accumulate a degree of wealth (and autonomy) and remained an exciting and invigorating prospect. For 70-year-old Bahadurer-ma, however, the migration of her sons was driven by poverty and, at the same time, did little to ease that poverty. Migration, in her case, was a form of distress diversification driven by enduring destitution.[3]

Mobility and livelihoods

Much social science work on migration has been negative in tone. To begin with, migration is seen as an outcome of poverty and underdevelopment (usually in rural areas). In other words, it arises out of an enduring failure to deliver better living conditions to marginal or peripheral areas and people. West Bengal is often thought to be just such an area, where 50 years of 'development' has failed to improve conditions for those most at risk with the result that there is a continuing migrant stream of the poor and marginalised to more prosperous Indian states, and to the slums of Delhi and other urban centres. In addition, the effects or impacts of migration have often been framed in negative terms. The process is seen to lead to the loss of human capital as the young, the able and the better educated leave home; it leads to deskilling as these trained migrants take up menial work which is beneath them; it compromises the social sustainability and coherence of communities as the 'left

BOX 6.1 The politics of mobility: China's *hukou* system

From the early Maoist period in China (the late 1950s), mobility has been tightly controlled by the household registration system or *hukou* which classifies households as either 'rural' or 'urban'. Not only were people often unable to leave their villages due to the systems of surveillance and control that existed at the time, but also the *hukou* also excluded households registered as rural from gaining access to a wide spectrum of socio-economic benefits and entitlements available to 'urban' households, even when those rural households might be living in urban areas. This helped to create and then to cement a deep divide between city and countryside by anchoring rural people in their rural place of residence. The *hukou* status of a family was generally passed to succeeding generations, fixing people in their geographical place.

The economic reforms of the 1980s, however, forced the Chinese state to revise the system because it was impeding economic growth and modernisation. Rather than simply permitting a mobility free-for-all, however, the state instead permitted 'temporary' migration of rural labour to urban areas through the awarding of temporary residence permits (*zanzhu zheng*). This so-called 'floating population' could number as many as 110 million workers and while they may be permitted to move under the new, revised *hukou* they continue to be denied many of the benefits available to urban registered households and are not eligible to take the more prestigious and better paid jobs available in urban areas. As Fan says, 'rural migrants are considered outsiders to the urban society, and most are relegated to the bottom rungs, picking up dirty, dangerous, and low-paying jobs and finding a marginalized and underclass existence in the city' (Fan 2002: 107). This division is not absolute; some well-connected rural migrants manage to negotiate a change in their *hukou* status and become, officially, 'urban' residents rather than belonging to the marginal and shifting 'floating population'. In addition, since 1992 the 'blue stamp *hukou*' has been another means by which some (privileged) rural migrants have managed to insinuate themselves into the urban fabric. The 'blue stamp' or *lanyin* awards those with valuable skills, education and money permanent urban residency and some of the welfare advantages that come from being an urban resident. Despite these changes to the *hukou* system, however, it remains a powerful means by which the Chinese state continues to act as a gatekeeper, controlling mobility through an internal passport system. 'In short, one's hukou or resident status continue to symbolize one's geographical (rural versus urban) origin, connotes one's socio-economic status, and above all defines one's opportunities and constraints' (Fan 2001: 485).

Sources: Fan 2001, 2002, 2003

behind' struggle to maintain the rhythms of life with unbalanced population profiles; and the presumed benefits of migration – in terms of remittances – is transitory and sometimes destructive as migrants and their families channel money into displays of conspicuous consumption.

Part of the reason for this negative take on migration and mobility can be linked to what has been termed the 'yeoman farmer fallacy' (Farrington et al. 2002: 15) and a normative position that many scholars, national officials and development practitioners take that rural people *should* remain in the countryside and, by association, in farming. 'It is when villages are fragmented by modernity, when village production is undermined by industrialisation, and when villagers are extracted from their natal homes that things are perceived to go wrong' (Rigg 2006: 187). Both Gill (2003) writing of rural Nepal and Deshingkar and Start (2003) on India lament the failure of policy makers to acknowledge the important and growing role of mobility in livelihoods. The latter, for instance, state that 'official awareness of the magnitude of seasonal migration or the importance of it in the lives of the [rural] poor is abysmally low' (Deshingkar and Start, 2003: 1). Rural people, as one might expect, are generally more sanguine in their views and pragmatic in their responses and see the opportunities available beyond the village and outside agriculture in a more positive light. 'Stay-at-home' policies and development strategies (de Haan 1999: 4) are often based on the premise that the movement is destructive and also that it is new and therefore an unwelcome change from established behaviours. But as De Neve says of migrants to the garment industry of Tirupur in Tamil Nadu, India:

> for many of the migrants, opportunities [for work] were abundant, promises soon materialised and expectations were fulfilled. In Tirupur, 'industrial modernity' turned out not to be a mere myth, but a reality which has fulfilled many expectations of migrant workers.
>
> (De Neve 2004: 277)

Scholars have come to recognise the importance of migration in livelihoods not just because they have challenged the yeoman farmer fallacy but also because studies have increasingly contextualised migration within the family and the household, rather than viewing the process as the outcome of decisions taken at the level of the individual (see de Haan 1999: 13–14). The household becomes the centre of analysis and migration, therefore, a process which is embedded in household social and economic relations, and part of a wider household livelihood 'strategy'. This was made clear in the discussion of Elmhirst's work in Lampung, Indonesia in Chapter 5 (see page 108).

Migration and livelihoods in India: a case study

A study of migration and mobility in twelve villages in the Indian states of Madhya Pradesh (MP) and Andhra Pradesh (AP) (Deshingkar and Start 2003) demonstrates two things: first, the degree to which migration is an important fact in the lives and existences of many householders; and second, the variation that exists in the levels of mobility between villages and between households (Figure 6.1).[4] Some of this variation can be regarded as 'old' and some 'new'. Take the case of the village with the highest levels of migration, MD in Medak, where 78 per cent of households have one or more members involved in migration. This village was one of the poorest in the sample, where local livelihood opportunities are limited by the dry, drought-prone conditions and local labouring opportunities are very scarce indeed, leaving most households with little choice but to engage in migration. Indeed, the average number of people migrating from each household was almost three, demonstrating the importance and ubiquity of the process. Large numbers of villagers in this area, and from similar contiguous areas of Karnataka and Maharashtra, migrate to Hyderabad, to other neighbouring states, and to the sugarcane fields of Andhra Pradesh. Moreover, these migration routes and streams are long established and, in that sense, 'old'.

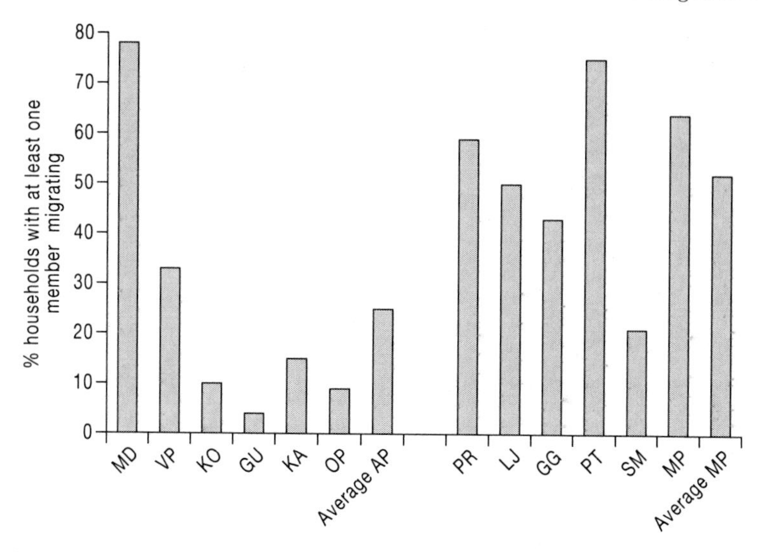

Figure 6.1 Incidence of migration in twelve villages in Andhra Pradesh and Madhya Pradesh, India

Even relatively resource rich and wealthy villages like KA and KO in Andhra Pradesh and SM and GG in Madhya Pradesh, however, have significant levels of migration, notwithstanding a relatively buoyant local labour market. This was because widening inequalities in access to land, spreading mechanisation, and unequal access to wage labouring opportunities were excluding a significant proportion of households from benefiting from these opportunities. Migration in the villages has emerged as a coping strategy in the context of a set of 'new' and emergent changes in the local economy, and a deepening class divide.

Not only did levels of migration vary, but also the contribution of migration to household income showed marked differences (Figure 6.2). This variation could not be easily explained. In Madhya Pradesh there was an inverse relationship between land ownership and the propensity to migrate – those with less land being more likely to be mobile. In Andhra Pradesh, meanwhile, there was no such relationship. This was partly because people were migrating – or not migrating – for different reasons and the aggregation of data was obscuring these differences. The very poorest often could not afford to migrate despite the fact that they needed to do so. Households with surplus labour were more likely to have a migrant household member than those in labour deficit. The research also indicated that Scheduled Tribes and Scheduled Castes (lower castes) were also more likely to migrate than the Forward Castes (Figure 6.3). But within this grouping there were some castes with skills to sell in the modern economy (e.g. the construction and brick-making skills of the Scheduled Caste Chamar) – who could access better paid work – while other backward castes were relegated to low return activities. Finally, while female migration was sufficiently widespread to challenge the view that it is largely men who move there were, nonetheless and again, marked differences between villages in the balance between male and female mobility. These differences can be linked to variations in local norms regarding the acceptability of female mobility, and women's skills and local gender divisions of labour (Figure 6.4).

In toto, this study from India demonstrates the centrality of mobility to the lives and livelihoods of many households in the study sites. It also shows how migration arises for contrasting reasons with the result that we cannot 'read off' from the fact of migration its root causes or, indeed, its likely effects.

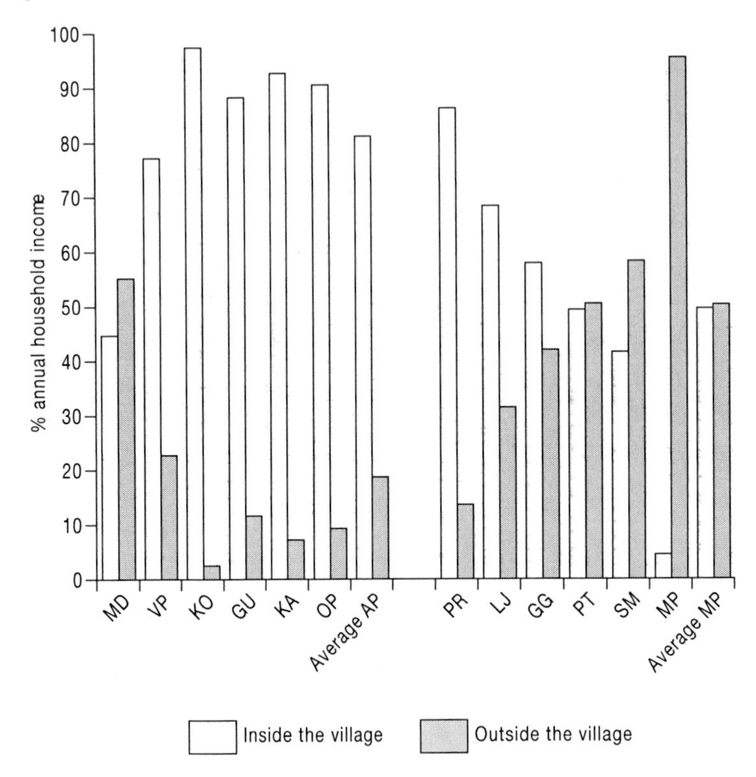

Figure 6.2 Returns to migration in twelve villages in Andhra Pradesh and Madhya Pradesh, India

Averages and particularities

Migrant experiences are necessarily mixed and while it is important to draw out general lessons from the diversity of experiences it is also necessary not to lose sight of the fact that these are 'just' generalisations:

> Migration may be propelled by poverty, and encouraged by wealth; it may reflect resource scarcities at the local level, or be an outcome of prosperity; it may be embedded in economic transformations, or better explained by social and cultural changes; it may narrow inequalities in source communities, or widen them; it may tighten the bonds of reciprocity between migrants and their natal households; or it may serve to loosen or break these bonds; it may help to support agricultural production; or it may be a means to break away from farming altogether.
>
> (Rigg 2007)

There are numerous axes that segment migration and migration streams: wealth/class, caste, ethnicity, religion, age (generation), gender, household size, marital status, occupational activity, education level, and skill acquisition, for instance. Furthermore, even after all these factors have been taken into account there remains a large black box which we can label 'unknown'. To put it another way, we should regard everyone in a given population as a potential migrant or as a possible 'stayer'. Neela Mukherjee's (2004) investigation into the lives and life chances of 60 elderly migrant women in Alaknanda slum in south Delhi, many

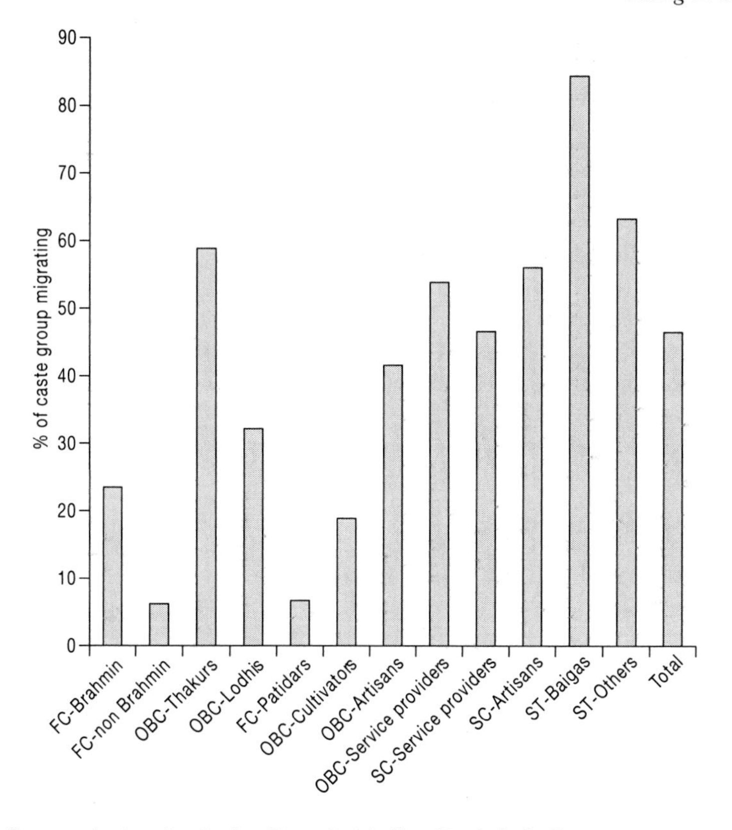

Figure 6.3 Caste and migration in six villages in Madhya Pradesh, India

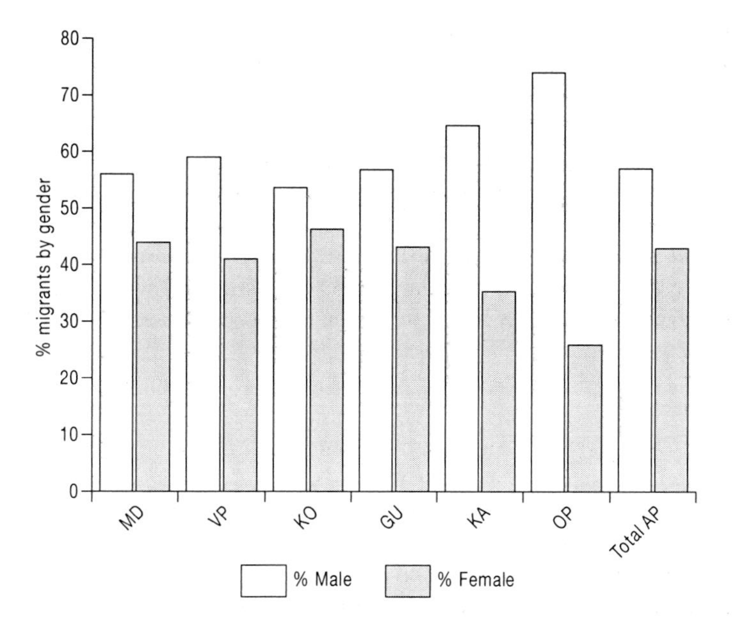

Figure 6.4 Gender and migration in six villages in Andhra Pradesh, India

of whom had left their impoverished families in West Bengal to seek alternative livelihood avenues, illustrates this tension between the general and the particular.

Rani left her husband and children in West Bengal because of 'abysmal' poverty. Her eldest daughter was put in charge of running the household in Rani's absence. Rani found work looking after an elderly woman in Delhi and managed to save and remit sufficient money to get her daughters married. She is now held in high esteem by her family and neighbours, she feels empowered, and she is in a position to help others with money. Sandhya, meanwhile, left her drunkard husband because he beat her, and fled to Delhi. She borrowed money to buy a hutment but because of a serious illness was unable to work and fell into arrears repaying the loan. At the time of the research she had still not fully recovered, was unable to work full-time, and was falling ever deeper into debt. Rani's experience illustrates that even when it is propelled by distress, migration may be financially liberating and socially empowering. Sandhya's experience, on the other hand, shows how the same process may lead the migrant into even deeper penury and greater vulnerability.

Migration as a social process

To summarise: people in the Global South are mobile, and increasingly so. It has also been shown, in the previous section, how there are tight links between livelihoods and mobility, to the extent that the livelihood logic is sometimes determining. Despite the generally welcome attempts to go beyond the economism that characterised much early migration research it would be wrong to ignore the fact that many individuals and families engage in migration for reasons that begin – and often end – as responses to livelihood threats, inadequacies or opportunities.[5] Generating income is regularly the undercurrent that drives migration streams. That said, migration is also – and importantly – a social process (Box 6.2) (see de Haan and Rogaly 2002: 6–9).

How migration is structured and organised is linked to social identities. In some cases, migration becomes a central component in how individuals think about themselves in the context of the wider community, and how societies collectively view themselves. In such cases, migration becomes an embedded and intrinsic, almost primordial, component of identity. Indeed some peoples' identities would seem to be defined less by the ways that they settle, but the ways that they move. This is clearest in work on Pacific islanders, migrants par excellence.

In their study of migration among the Wosera Abelam of Papua New Guinea, Curry and Koczberski (1998: 36) argue that 'a migrant never really leaves the village in a spiritual and cultural sense, and hence there is no clear cut distinction between sources and destination sites when examining the lived experience of Wosera migration'. They tentatively suggest that the destination site is an extension of home (1998: 47). This resonates with Chapman's (1995) work on Melanesia in which he quotes John Waiko, who explains: 'We Melanesians are all engaged in a tapestry of life, where the threads of movement hold everything together' (Chapman 1995: 257). Mobility, for some Pacific Islanders, is so central to their lives that to equate place of residence with home is to ignore this defining characteristic.[6]

In a second article on longer term migration (rather than circulation) among the Wosera Abelam, Curry and Koczberski (1999: 140) note that even when migrants have been absent for fifteen years or more they retain a 'strong ideology of return'. In this article, however, there is a more pronounced sense of 'home' and 'away' and while migrants continue to engage in indigenous exchange activities this is seen as a means of legitimising their claims to resources in the natal village. There is also evidence that the maintenance of social identity as a means of protecting traditional rights is coming under pressure 'and may no longer be

BOX 6.2 Modernising and liberating Islam: migrant women from Sri Lanka to the Middle East

It is sometimes argued that migration, and the engagement of migrants with a social world beyond their own, can loosen and even rework social relations at home. Migration becomes a force for social change, and a liberating and empowering influence. Thangarajah's (2004) work among female migrants from Muslim communities in eastern Sri Lanka to the Middle East shows something rather different and demonstrates the need to entertain the possibility of sharply varied outcomes from similar processes.

Women from Muslim communities in eastern Sri Lanka began to migrate to take up work as domestic servants in the households of well-to-do Arab families in countries such as Saudi Arabia, Oman, Dubai and Abu Dhabi in the mid-1980s. Drawing on his work in the village of Ullur in Batticaloa district, Thangarajah (2004) shows how these women are drawn into a cultural and religious context very different from their own almost as soon as they step off the plane. After verifying that the women are, indeed, Muslim, they are taken to the *souke* and a *hijab* (*abbhai*) purchased for them to wear. They quickly make unfavourable comparisons between religious practices at home and those in their country of work, let alone the standards of living in each. 'Arab' religious practices are not regarded as 'backward', far from it; they become markers of sophistication and avenues for the accumulation of social respect. Thangarajah sees altered religious practices becoming dovetailed with higher levels of consumption and an engagement with a high-tech, modern experience. Religion thereby becomes, by association, empowering, liberating and modernising. These women return to become (or resume the role of) mothers, but do so with bolstered authority, a higher status and a burnished image, not to mention greater wealth: 'both Islamic orthodoxy and domesticity, generally seen as restricting for women, instead opened up a space of social mobility, financial independence and empowerment' (Thangarajah 2004: 149). (Silvey's (2004) sharply critical work on Indonesian female migrants to Saudi Arabia provides a valuable counterpoint to Thangarajah's study (see page 161).)

These female migrants use their experience of working in the Middle East to rework social relations at home in ways that accentuate seclusion and construct a new Islamic identity. Money accumulated abroad is used, first and foremost, to build a house, as has been shown to be true of international labour migrants in many other countries. But these buildings incorporate not just new notions of what it is to be modern in the secular sense – attached toilet facilities for example – but also new religious beliefs. Public and private space is clearly separated as is men's and women's space. In terms of clothing, female migrants often choose to continue to wear the hijab although Thangarajah (2004: 157) argues that this is 'a sign of confidence, high fashion and economic power'.

appropriate nor valid in contemporary Wosera society where the rules governing social organisation and resource access are tightening in response to growing population pressure' (Curry and Koczberski 1999: 142). Two key changes are arising from escalating resource pressures in sites of origin. First, mobility patterns are being reworked as formerly dominant short-term circulation is superseded by longer-term – possibly permanent – migration. Second, the links between migrants and their natal villages are undergoing change as migrants' traditional rights, maintained through ongoing contact and engagement with home, are compromised. It is tempting to think that, in the longer term, these changes will have implications for Melanesians' view of migration, of 'home' and 'away', and will therefore also lead to a shift in the academic view of migration and mobility among such peoples.

We should be careful not to see migration simply as a 'response' to household and local conditions, whether those are framed in economic or socio-cultural terms. The danger of writing of 'distress migration' or 'migration for accumulation' is that mobility comes to be seen as a means to an end, or as an outcome of a set of structures and relationships that have a degree of fixity. It is also important to see the process itself – migration – as a means by which these very structures and relationships can be changed. Thus it has been argued that the migration of women in South Asia has led to a 'redefinition' of motherhood: female migrants nurture strangers' children away from home, while earning and remitting money to support their own children, who are being raised by others (Gardner and Osella 2004: xxiii). Furthermore, the experience of migration forges a sense of collective identity among migrant populations, creating a bond between sometimes disparate groups and individuals who 'at home' would have little reason to cooperate or find common cause. It is therefore necessary to look 'forward' to destination sites, as well as 'back' to source areas.

Yan Hairong's (2003) study of female migration from rural villages in Wuwei County, Anhui Province, China demonstrates the power of migration to remake identities and to metamorphose over time. Wuwei County has become known as a source of female domestic workers, something which dates back to the Revolution when, it is said, some army officers left their children to be cared for by women during the war with the Republican forces. Following the founding of the People's Republic in 1949 some of these women were offered employment as domestics in the households of veterans of the Revolution. By 1970 some 3,000–4,000 women from Wuwei had worked as domestics in the larger cities of China and this figure continued to grow so that in 1993 the total number of rural Wuwei women taking on such work was 263,000 (Yan Hairong 2003: 579–580). Two things stand out about migration during this early period. First, women continued to see themselves – and to claim – a rural identity. They were *nongcun funü* – rural women – even though they lived in the city. Second, such work was regarded as slightly embarrassing or unseemly, a source of shame. For this reason, until the 1980s, young women did not migrate; to do so would have undermined their marriage prospects. Through migrating, rural women, until the 1980s, can be thought of as transgressors in two senses. To begin with, because they transgressed the norms dictated by local, rural patriarchy; and second, because they transgressed the revolutionary ideals of the heroic, female farm labourer, 'invoking the specter of the [feudal] past through domestic service' (Yan Hairong 2003: 582). Furthermore, and as a result, the money earned from this work was somehow 'unclean'.

Beginning in the 1980s, migration from Wuwei began to change. Young women migrants began to leave the countryside to take up urban work, and to do so in their hundreds of thousands. The migrants of the 1960s and 1970s were intent on survival; those of the 1980s and 1990s, on wealth and the acquisition of the material and non-material elements of modernity. In a few years between the Maoist and post-Maoist periods, the countryside became transformed from a socialist site of revolutionary progress, heroism and fervour to modernity's Other, a backwater where nothing happened and where prospects were dim. From migration itself being an embarrassment in the 1970s, the countryside had become an embarrassment and migration the means of escape. 'I don't want to be like my parents – their life [in the countryside] is going nowhere. . . . What matters to me is that I want to have some achievement. I want to live like a human [*ren*]' (Yan Hairong 2003: 587). Staying in the countryside for a young woman means rural domestication and the curtailment of any possibility of accessing and achieving modernity.

Exode *among the Fulani of Burkina Faso*

The interplay of culture and economy can be seen in migration patterns among the Fulani of the Sahel. The Fulani are highly mobile.[7] Hampshire (2002) surveyed 834 households in 40 Fulani villages in northern Burkina Faso in 1995, with a follow-up qualitative survey among a sub-sample of 6 villages between 1995 and 1996. The study focused particularly on one form of mobility: temporary dry season migration to urban centres outside the local region, which Hampshire calls, using the French word, *exode*. *Exode* appears to have first emerged in 1973 during the course of a serious drought, which acted as a trigger point. In economic terms it is clear that this form of mobility is biased in favour of resource-rich households (Figure 6.5) and those households that have more adult men to release for migration.[8] This pattern can be explained and understood in economic terms. *Exode* is expensive, demanding a considerable initial outlay. Poorer households have other, less risky, coping options open to them which are not so asset-intensive. Furthermore, the returns to *exode* are lower for poor migrants because they tend to beg or find work in the informal sector, both highly marginal activities in the urban context. Better-off migrants, by contrast, tend to engage in activities such as cattle trading and various forms of self-employment for which returns (and risks) are higher. 'In other words', Hampshire (2002: 23) concludes, 'wealthier households are not only better able to send migrants *per se*, they are in a better position to access the most profitable forms of *exode* which can make a substantial contribution to livelihood security.'

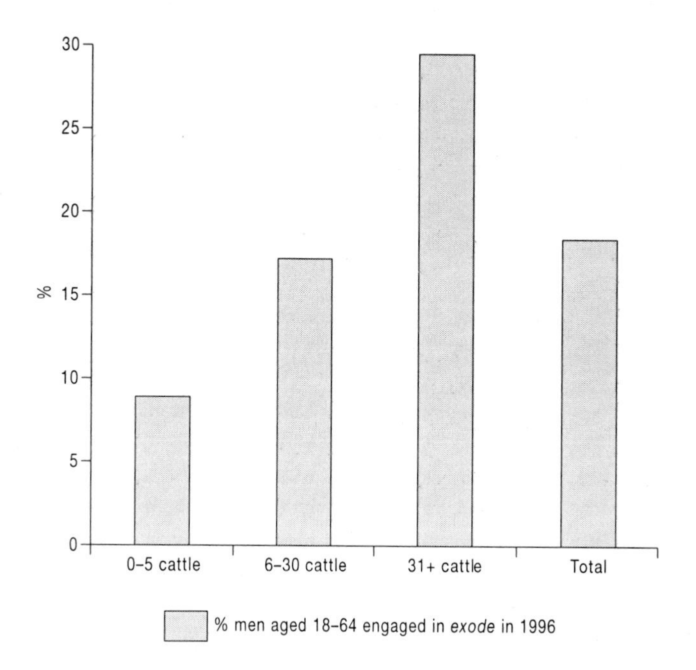

Figure 6.5 Migration and wealth among the Fulani of Burkina Faso

But for Hampshire (2002), seeing migration in economic terms supplies only part of the necessary explanatory furniture. This is where social process and identity come into play (Table 6.3). Of the 346 migrants Hampshire surveyed in 1994–1995, 333 were male. Strictures on women's mobility and on their activities were strong and *exode* emerged within and was moulded by this pre-existing structure of gender relations.[9] Social networks also proved to be important in securing work and accommodation in the destination area for male migrants and in filling any labour deficits at home following a migrant's departure. A third, very important thread of explanation is linked to contrasting ethnic sub-group identities. The *FulBe GaoBe* showed a very clear propensity to migrate compared with other ethnic sub-groups; for the *FulBe DjelgoBe*, by contrast, *exode* was rare and regarded as degrading because it challenged and undermined their identity, first and foremost, as cattle herders and pastoralists. Furthermore, the type of work undertaken was also quite sharply segmented by ethnic sub-group. Finally, there was a generational dimension to *exode*. Men in the 28–40 age group were the most likely to migrate. Hampshire (2002) suggests that this was because these men have grown up with *exode* and it does not hold the shameful connotations that are evident among the older generations.[10]

Hampshire (2002) provides a convincing account of how we can see patterns of migration among the Fulani of northern Burkina Faso as linked to social and economic relations centred on class (wealth), ethnicity, gender and generation. All these axes of explanation, however, also have to be viewed in historical context. Identities and accepted practices are being continually negotiated and renegotiated so that one generation's migration practices will not be handed down unaltered to the next (Box 6.3). As is outlined below in more detail, migration is not just a response to or a reflection of such relations and structures; migration and mobility can also create those relations and structures.

Table 6.3 Exode among the Fulani of Burkina Faso: axes of explanation

Axes of explanation	*Evidence*
Economic	
Wealth	Direct and statistically significant relationship between migration and cattle holdings (see Figure 6.5)
Resources	Men from larger households and households with larger numbers of adult men are more likely to engage in exode
Social	
Gender	Just 4% of migrants (n = 346) were female; local norms and accepted practice largely precluded women's involvement in *exode*
Social networks	Migrants used social networks to secure work and find accommodation in the urban centre while migrant households used such networks to fill any labour deficits that might emerge following the loss of the male migrant
Ethnicity (Fulani sub-group)	Statistically significant link between migration and ethnic sub-group reflecting contrasting sub-group identities with the *FulBe GaoBe* migrating and the *FulBe DjelgoBe* avoiding *exode*
Generation	*Exode* most common in the 28–40 age group because these men have grown up with the reality of migration as a livelihood option and are therefore less resistant to embracing it

Source: information extracted from Hampshire 2002

BOX 6.3 From *Orang Kampung* to *Melayu Baru*: evolving migrant identities in Malaysia

Thompson (2003: 418) makes a case for the 'dissociation' of the Malay-peasant complex and does this by reference to the life histories of two young men from the village of Sungai Siputeh introduced at the end of Chapter 5 (see page 115). Idris is a working class male; Nizam is middle class. For Nizam the experience of working in a managerial position in an urban context had driven a wedge between him and his place of origin. But this divide was neither sharp nor absolute. His home village had become both an idealised place where he could seek sanctuary from Kuala Lumpur *and* a source of some embarrassment, a backward place where people held simple, narrow views. Idris, by contrast, moved between his urban place of work and his village of birth with greater ease. Migration had led to a dissociation between these two men and their former identity as *Orang Kampung* (Village People), but differently:

> their experience of migrancy took place on a terrain intersected by fields of class and other identities. For Idris, the experience was one of broad horizons limited by his modest accomplishments in primary and high school. . . . Nizam, on the other hand, was able to pick and choose among a variety of possibilities.
>
> (Thompson 2003: 424)

For both Idris and Nizam, migration had played an important role (although it was not the only influence) in taking these two men along the multiple paths from *Orang Kampung* to *Melayu Baru* (New Malay). Working class Idris may never become a New Malay in the Malaysian government's image; but even so, the experience of migration has meant that he has been directly caught up in the unfolding of modern Malaysia.

Migrant identities and the politics of difference

While migration can lead to the blurring of boundaries as contact with other groups leads to a process of acculturation, it may also serve to harden distinctions and divisions as migrants forge stronger collective identities, and 'receiving' destinations develop particular views and positions regarding migrant populations.[11] This sharpening of difference is seen, for example, in the intensifying conflicts between migrant settlers and established landowners in oil palm settlement areas in West New Britain Province (WNBP) and Popondetta in Papua New Guinea:

> In both WNBP and Popondetta, ethno-regional identities are forming among landowners that have one's ethnicity, province of birth and claims to ancestral land as the primary markers for inclusion. Land, place and belonging are emphasised. Among migrants in WNBP, a shared identity is gathering strength based, not on common descent, but on shared experiences, insecurities and a constructed history of their role in national development.
>
> (Koczberski and Curry 2004: 362)

At one level these sharpening conflicts are ethnic. At the same time, though, they are about livelihoods and about the ways in which contemporary political and economic change has reworked livelihoods and access to livelihoods in ways that are perceived to benefit one group, or groups, over another or others. 'Landowners', as a class, do not share a common culture or language. There is little in historical terms that might bind what would seem to be disparate groups. But a sense that they – as a group – had lost control of 'their' land to migrants was creating a sense of common grievance and therefore of common purpose and common identity. Moreover, the politicisation of these grievances was sharpening this sense of common cause and deepening the divide between landowners and migrants. A similar process has occurred among the multi-ethnic migrant population where recent events have encouraged the emergence of an identity which cuts across (but has not yet replaced) difference. It is international migration, however, where constructions of difference are often most pronounced. In many cases migrants are living and working not only in a new country, but also in a new cultural context where there are marked differences of language, dress, religion, diet and habit.

To fuel its rapid economic expansion, Malaysia has seen the long-distance migration of hundreds of thousands of workers from Indonesia, Burma, the Philippines, Thailand, Nepal – and from Bangladesh (Illustration 6.2). In 2005 it was reported that there were 2.6 million legal and illegal foreign workers in Malaysia in a total workforce of some 10.5 million.[12] Malaysia itself is a multi-ethnic society and yet Bangladeshi migrant workers have become cast as the 'Others' in their host country, and this despite the 'Muslim Brotherhood' that was meant to bind Bangladeshi workers with the majority Malays (Dannecker 2005: 251). In 2001

Illustration 6.2 A migrant worker recruitment and training agency in Penang, Malaysia

there were between 100,000 and 300,000 Bangladeshis working in Malaysia in fast food restaurants, on petrol station forecourts, in the construction, manufacturing and plantation sectors, and in domestic service. The axis of difference in Malaysia has traditionally been drawn between the three main communities that make up the Malaysian multi-ethnic state: Malays, Chinese and Indians. Yet the burgeoning population of migrant workers in Malaysia has caused this to change so that, and increasingly, the 'Others' have become migrants – irrespective of their national, cultural or religious affiliations. The media have assisted in the process by writing of migrants as if they are a single group. Criminal acts committed by individual migrants cause all migrants to become labelled and seen as a generalised threat to Malaysian society. This has, in turn, created a sense of a Malaysian collective society which transcends traditional lines of difference. Bangladeshi migrants have become particularly demonised in the public consciousness as, variously, a threat to local womanhood (because they court local women), because they take jobs from local people (particularly Malaysian Indians), and because they compete for and force up the price of low cost housing (Dannecker 2005). The presence of migrants in Malaysia has, as in Papua New Guinea, encouraged a multi-ethnic host society to forge a greater sense of collective identity. At the same time, migrants facing media stereotyping, popular resentment and legislative exclusion have also found common cause.[13]

Migration, source communities and the 'left behind'

The relationship between conditions in source communities (places of origin), patterns of migration, and the impacts of mobility on those source communities and the 'left behind' is a complex one. The circular migration of males from the Indian state of Bihar, for example, has persisted for more than a century (de Haan 2002: 115). Bihar today is one of India's poorest states but migration predates its stagnation relative to the rest of the country. So while circular labour migration by males may have been a feature of Bihar from the late nineteenth century the form that migration has taken – for example in terms of destination and work available – and the motivating forces underpinning the process have varied significantly over time. The tendency in the past for scholars to focus on the migration process itself has tended to distract attention from context and effect.

Just as migrant identities are changed by the experience of moving, so too with rural source contexts more generally. In Chapter 5 it was suggested that the engagement of young women from rural Lampung in factory work around Jakarta had significant social as well as economic ramifications for their home village of Tiuh Baru.[14] Sen (1999) goes rather further in her study of the nineteenth century colonial jute industry in India where the selective migration of men to the jute mills of Bengal has served to shape – rather than just be a response to – gender relations and gendered divisions of labour. Indeed we can push this explanatory envelope a little further. It is possible to argue that social relations (gender, generational, class, and more) in rural *and* urban areas, in agriculture *and* industry, in source *and* destination areas, in capitalist *and* pre-capitalist systems are co-constituted and that the separation of these spatial and sectoral arenas overlooks such links and associations. We should be looking, in other words, at the totality of the system in which mobility is a component part, rather than examining the rural and urban contexts as two, largely discrete systems linked by migrant flows.

Mobility and structural change in the rice basket of Java

The Indonesian island of Java supports some of the most densely populated rural areas in the world. During the 1970s and 1980s, under the twin pressures of population growth and the wealth-concentrating effects of modernisation, many scholars expected there to be a mass exodus of rural people to urban centres as poor households were squeezed and then displaced from the countryside. This mass of rural migrants to the city would then form an urban proletariat, struggling and disenchanted, and represent for the Indonesian government both a humanitarian challenge and a political worry. Overall, such a scenario has not transpired. Rural populations have continued to grow, land holdings have continued to shrink, and agriculture has continued to advance in technological terms; but rural households have, at the same time, continued to persist seemingly against the grain and the tide of demographic, technological and resource trends in the countryside. To understand this surprising resilience of rural Java's human landscape we need to look at the central role that migration has played.

North Subang in West Java has been intensively studied since the 1970s by a succession of scholars from a range of disciplines.[15] It is situated in what is termed the 'rice basket' of Java and, in visual terms, continues to appear to be an agrarian community (Breman and Wiradi 2002: 42) (Illustration 6.3). But behind the appearance of agrarian continuity, North Subang has been undergoing dramatic change. Between 1990 and 1998 the share of the working population engaged in agricultural activities dropped from 75 per cent to 58 per cent. By 1998, two-thirds of households had at least one member working fully or largely outside agriculture, up from one-third in 1990. The changes to the composition of North Subang's economy are tied to the growth of extra-local, non-farm employment. In 1990, 40 people worked outside the village; in 1998, the figure was 82. These 82 villagers represent a diverse group working outside North Subang for a range of reasons, in a range of occupations, and with a range of possible livelihood outcomes (Table 6.4).

What does this mean, though, for the nature and functioning of North Subang, as a community and how does it link with the notion of sectors and spaces being co-constituted? To begin with, the image of rural continuity that North Subang presents, perversely, has been maintained only because things have changed.[16] The dislocation of rural communities that was predicted in the 1980s has not occurred because of the way that mobility has permitted villagers to access opportunities outside the local area. The income generated has allowed households with sub-livelihood land holdings to survive in the village even if they do so by engaging with work beyond the village, and outside farming. It is not just North Subang's economy and livelihoods which have to be contextualised across space and between sectors; changing social identities are also a product of the same evolving interdependencies. For those with regular employment (Type I in Table 6.4), the link with farming has been severed and when these men and women return to the village they do so carrying the baggage of urban life and work, and the aspirations that inform them. Their value to those who remain in North Subang is to act as conduits or 'bridgeheads' (Breman and Wiradi 2002: 113) to urban work opportunities. For the workers in the Type II and Type III categories in Table 6.4 the break with agricultural work and their separation from the village as an economic and social unit is not so abrupt. For these individuals it is the *combination* of work across spaces and sectors that delivers resilience. They cannot depend on intermittent, marginal and low-paid work in the urban context for their livelihoods; nor can they rely on rural-based livelihoods when land is scarce and other local opportunities are few and deliver inadequate returns.

Through the lens of the interlocking livelihoods perspective, we give migration an economic logic which would seem to deliver a degree of continuity. But, as previously noted,

Illustration 6.3
The 'rice basket' of Java, Indonesia

migration is also a social process and this can serve to undermine its economic validity. Koning (2005) entitles an article on young rural women migrants to urban Jakarta, 'The impossible return?' As with North Subang, her village of Rikmokèri in Central Java was one where migration represented a critical pillar in the village economy with 50 per cent of 688 households having a family member working in Jakarta. She asks: what happens when young women migrants return to the village? 'Once they have experienced urban life, many daughters feel caught between two worlds: between village and city; between childhood and adulthood; between expectation and reality; between their own aspirations and what their parents expect of them' (Koning 2005: 169). Her respondents struggle against the roles that are created for them and the expectations that are required of them in the village. She argues that the measure of a person in the rural context is tied to place-based and time-dependent factors, 'of the owner of so much land, of that specific house, or belonging to that family' (Koning 2005: 171). In the urban context it is what people do that counts and many women migrants resist being drawn back into the village. Breman and Wiradi's (2002) work

Table 6.4 Patterns of extra-local, non-farm employment in North Subang, Java, Indonesia, 1998

Employment/occupation	*Number*	*Livelihood outcome*
Type I: Permanent or regular employment	9 (7 men, 2 women)	Skilled and well paid, permitting some savings and the accumulation of capital
Factory workers	3	
Shop assistant	2	
Drivers	2	
Others	2	
Type II: Self-employed, mostly in Jakarta	19 (15 men, 4 women)	Generally low return, with some possibility for upward mobility
Sinkong (fried cassava) sellers	9	
Prostitutes in north Jakarta	4	
Hawkers of cigarettes, newspapers and other items on buses and in markets	6	
Type III: Casual, itinerant labourers	54 (45 men, 9 women)	Unskilled, low paid and intermittent work with little scope for saving or upward mobility
Casual earth movers or *kenek* (navvies) for the building sector	45	
House maids, street stall assistants	9	
Total	**82**	

Source: information extracted from Breman and Wiradi 2002: 111–116

emphasises the degree to which the vitality of North Subang is dependent on migration and extra-local work; Koning's research in Rikmokèri reveals how the experience of migration for the migrant may, over time, drive a wedge between their aspirations and desires and those of their families and the wider community.[17]

Sometimes it is necessary to return to the mundane to ferret out the answer, because occasionally the mundane can tell us as much as the apparently profound. In 1990, North Subang was accessible only along a poor track that took 45 minutes to negotiate on foot. By 1998 the track had been asphalted. In 1990, North Subang appeared 'backward and inaccessible'. In 1998 it was highly connected in a myriad of ways: by surfaced road; by electricity; by TV; and by telephone. This change from the (apparently) disconnected life of the peasant to the connected life of the post-peasant was, moreover, concertinaed into less than a decade. Embedded in and arising from these physical changes are the social and economic adaptations and transformations noted in the foregoing paragraphs. It is all too easy to overlook the materiality of connection because it seems just so very obvious – and perhaps just a little bit too matter-of-fact to be a subject of concern for high-level scholarship. It is, nonetheless, often the elephant in the room of explanation.

'Connection', of course, is a mental as well as a physical process. Migration brings new flows of ideas, wishes and desires, as well as money, which can – and do – rework locally imagined futures. As McKay (2003; see also McKay 2005) shows in the context of the Ifugao of the Philippines, these re-reworked futures feed back into decisions that are taken regarding, for example, what crops to grow:

> the crops that are planted in Ifugao fields say volumes about how the people planting them envision themselves in relation to both the state and to global labour markets. Bean gardens can be read as remittance landscapes – they both anticipate remittances and produce the capital needed to go overseas – and are thus tied to the translocal nature of apparently local places.
>
> (McKay 2003: 306)

Migration, mobility and source communities: for better, for worse?

The economic and social impacts of migration on the 'left behind' have been debated for some time. This is partly because many politicians and government officials feel that they should have a 'position' on migration if only so that they can decide whether it should be controlled, prevented, managed, encouraged or, simply, ignored. This creates the temptation to generalise about migration on the grounds that what appears to be a uniform process should have uniform causes, uniform outcomes and, therefore, uniform lessons. Views on migration arise out of the building of a series of generalisations:

- If we accept that migration which permits a degree of accumulation (i.e. it is prosperity-raising and asset-forming) is generally the preserve of the middle and rich, then mobility will tend to widen inequalities in source regions, creating a deeper divide between asset poor and socially excluded non-movers, and relatively richer movers (Waddington and Sebates-Wheeler 2003).
- If we accept that, traditionally, villagers were sedentary then migration necessarily disturbs the status quo, upsetting the social and economic quiescence of the countryside.
- If we accept that the bulk of remittances are either ploughed into performances of conspicuous consumption or force up the price of rural assets (particularly land) then migration will bring tension, rather than progress.
- If we accept that migration is gender selective then it will deliver either cumulative benefits or cumulative costs to men or women. In rural Nepal, most migrants are male. With their husbands or fathers away, female stay-at-homes find that the responsibility for subsistence production falls on their shoulders, with a resultant increase in women's (already high) work loads, a feminisation of farming, and a marginalisation of women in agricultural work (Gill 2003: 27).

On the basis of these sorts of generalisations it is easy to see how officials arrive at the view that migration is problematic and should be controlled. However, such generalisations only operate for some people, in some places, with regard to some issues, and at some points in time/history. The discussion in this chapter has shown how we need to put some intellectual space between the identification of mobility/migration as a process, and its interpretation.

There is one further point to make. For many, whether or not to migrate is not an option or a lifestyle choice. It is an economic and social necessity. As the discussion of North Subang shows, there is little scope of returning to an idealised sedentarist past. Individuals and communities are adapted to livelihoods which cross spaces and sectors. The entrenched view that during times of crisis source communities will provide a safety net is, for this reason, questionable. During the Asian economic crisis of 1997–1998 some commentators and activists expressed the view that the downturn would provide an opportunity to reinvigorate rural areas and communities (see Rigg 2002, 2003: 112–113). Certainly, many millions of workers were laid off. It is also true that many millions in Thailand and Indonesia returned 'home'. But, as Silvey (2001) concludes in her work in Indonesia, migrants left her research site of Maros in South Sulawesi in part because the local economy did not provide adequate support. Furthermore, over time these households have adapted to and become dependent on remittances from migrant family members. There might be a degree of romantic, populist appeal in seeing the countryside 'saving the day' when modern lives are compromised but the reality is that, in many cases, mobility's embrace is such that the traditional/rural/agricultural and the modern/urban/non-agricultural are intimately joined through their co-constitution.

Turning the tables: cities as source communities

The wider context and processes that inform this chapter are increasing rates of urbanisation and growing levels of rural-to-urban migration. 'Source' communities are, therefore, almost inevitably rural. There are, however, some examples – particularly from Africa – of counter-urbanisation or 'reverse migration' where cities have become source 'communities'.

Zambia has been experiencing a process of counter-urbanisation since the 1980s. In 1980, 40 per cent of the population was urban; in 1990 the figure was 39 per cent; and in 2000, 36 per cent – a trend that Potts describes as 'astonishing' (Potts 2005: 590).[18] In short, since the early 1980s, Zambia has been ruralising. While the process of counter-urbanisation in the West is associated with affluence and a desire among people in urban areas to access the perceived better living conditions in rural and suburban areas, in Zambia it has been associated with economic stagnation, urban unemployment, declining real urban incomes and standards of living and, therefore, the need to access rural areas for survival. Furthermore, it seems that this out-migration not only is restricted to those who were born in rural areas but also includes those who were born in urban centres but have been forced out for economic/livelihood reasons.

This surprising trend in Zambia is also repeated in other African countries such as Zimbabwe, Tanzania and Ghana where urbanisation rates have either slowed in a manner that was not anticipated, or have reversed.[19] Counter-urbanisation in Africa is not, usually, driven by choice, but by necessity. As formal employment has evaporated, real incomes have declined, and subsidised urban services have disappeared so urban residents have embarked on new or adapted livelihood strategies to keep their heads above water. The informal sector has blossomed, urban agriculture has expanded, and occupational multiplicity – the juggling of several occupations – has spread (see Potts 1995, 2000). In some cases circular migration has replaced more permanent relocation as people try to keep a livelihood 'foot' in urban and rural areas. Inevitably, however, these adaptive strategies have not always been sufficient, and many urbanites have had to take the final solution and 'return' to rural areas. Usually this has been a case of 'going home'; whether it will mean 'staying home' is to be seen.

To highlight a point made in the opening chapter of the book – that we should not expect similar processes in the Global South and Global North to have the same underlying logic (see page 4) – the interpretations that have been applied to counter-urbanisation in Africa (as above) are very different from those that inform the process in Europe.

Conclusion: returning to the new mobilities paradigms

The socio-economic and demographic conditions and circumstances that we find across the Global South not only provide the explanatory context *for* migration, but also represent a context arising *from* migration.[20] To put it another way, migration streams arise or are bolstered due to the forces of modernisation; they also arise or are accentuated because migrants want to become modern. This chapter has largely concerned itself with migration as viewed in the context of source rather than destination communities. However the other 'end' of the migration process has been an object of interest in earlier chapters, for example with regard to rickshaw pullers in Dhaka, Bangladesh (see page 84) and ideologies of modernity in Zambia (see page 65). It is necessary for academics to be as connected in their consideration of migration as the migrants themselves and, to date, this has been comparatively rare. We find it hard to escape from the categories that confine and define our thinking.

McKay (2003, 2005) writes of 'remittance landscapes' in the context of land use decisions among the Ifugao of the Philippines (see above) where it is possible to 'read' remittances and overseas contract work not only into gender relations in the study area but also into the very landscapes of the region. It is useful to widen this idea to a consideration of 'migration landscapes' in more general terms. What we see in the villages and fields of the Philippines, India and Indonesia, and in the cities too, are social, economic and physical landscapes which are being refashioned by migration. What people do, how they do those things, how people relate to each other, what and where they eat, and the aspirations they hold dear, for example, are either outcomes of migration or are affected by the process.

This chapter began with a consideration of what a 'Southern new mobilities paradigm' might look like, and set that against a Northern new mobilities paradigm (see Table 6.1). To complete the chapter it is worth returning to the table to consider some of the broader issues that it raises in the context of the discussion.

There are four areas where a Southern focus on mobility and migration offers something new and different from much of the material which draws on the Northern experience for its empirical inspiration. These are:

- migration, connection and the co-constitution of places, people and activities
- the balance and tension between the individual, family/household and the village/community as units of study
- the links between migration and mobility and livelihoods
- the essential ambiguity of migration.

These differences are linked both to the approach and viewpoint that many scholars working in the Global South adopt, and the material and evidence that they glean. So it is not just what they find, but also how they look.

The Western Isles of Scotland have experienced years of rural out-migration as the young and the educated have left their homes and escaped to other areas of Scotland (and beyond) offering better employment opportunities and more attractive lifestyles. The source communities, for their part, have been left with ageing populations and generally low education and skills levels. For Stockdale, 'those with either academic ability or ambition appear to have little choice but to leave the rural community' (Stockdale 2004: 188; see also Stockdale 2006). Individual migrants gain from the decision to move away; communities lose and rural society collapses. On the face of it, the experience of communities in the remote Western Isles of Scotland is similar to that of many rural areas of the Global South. The critical difference is that in most communities in the Global South the natal bond that links the migrant and his or her household has not been severed. In consequence, there is a much greater need to see source and destination areas, and migrants and non-migrants as intimately linked in both material and non-material ways. It also means that livelihood outcomes are closely associated. Finally, it requires a careful juggling not only of scale (local/extra-local, rural/urban) but also of the units of analysis – individual, household, family, community – that we apply.

Further reading

Since the early 1990s there has been a growing recognition of the central role that migration can play in livelihoods. The articles by de Haan (1999) and de Haan and Rogaly (2002) summarise some of the wider themes while Deshingkar and Start (2003) provide an assessment based on work in India. Migration may have an underpinning economic/livelihood logic but it usually leads to social change, sometimes profound. Gardner and Osella (2004) give an overview of such transformations and the case studies in their edited volume detail the situation in different contexts across South Asia. Migration can also lead to growing antagonisms between migrants and their host communities, as Dannecker (2005) shows in her study of Bangladeshi labour migrants in Kuala Lumpur and Koczberski and Curry (2004) in their work on settlers in Papua New Guinea. For a while there was a tendency to focus on migrants, and to ignore the effects of migration on source communities and the left behind. Koning's (2005) work on Indonesia and McKay's (2003, 2005) on the Philippines show how migration changes not just the migrants but their families and villages of origin as well. A useful corrective to the usual assumption that flows are rural-to-urban can be found in Potts' (1995, 2005) studies of counter-urbanisation in Africa.

Migration and livelihoods

de Haan, Arjan (1999) 'Livelihoods and poverty: the role of migration – a critical review of the migration literature', *Journal of Development Studies* 36(2): 1–47.

de Haan, Arjan and Rogaly, Ben (2002) 'Introduction: migrant workers and their role in rural change', *Journal of Development Studies* 38(5): 1–14.

Deshingkar, Priya and Start, Daniel (2003) *Seasonal migration for livelihoods in India: coping, accumulation and exclusion*, Working Paper 220, London: Overseas Development Institute. Downloaded from http://www.odi.org.uk/publications/working_papers/wp220.pdf.

Migration as a social process

Gardner, Katy and Osella, Filippo (2004) 'Migration, modernity and social transformation in South Asia: an introduction', in: Katy Gardner and Filippo Osella (eds) *Migration, modernity and social transformation in South Asia*, Contributions to Indian Sociology, Occasional Studies 11, New Delhi: Sage, pp. xi–xlviii.

Migration and the politics of difference

Dannecker, Petra (2005) 'Bangladeshi migrant workers in Malaysia: the construction of the "Others" in a multi-ethnic context', *Asian Journal of Social Science* 33(2): 246–267.

Koczberski, Gina and Curry, George (2004) 'Divided communities and contested landscapes: mobility, development and shifting identities in migrant destination sites in Papua New Guinea', *Asia Pacific Viewpoint* 45(3): 357–371.

Migration, source communities and the left behind

Koning, Juliette (2005) 'The impossible return? The post-migration narratives of young women in rural Java', *Asian Journal of Social Science* 33(2): 165–185.
McKay, Deidre (2003) 'Cultivating new local futures: remittance economies and land-use patterns in Ifugao, Philippines', *Journal of Southeast Asian Studies* 34(2): 285–306.
McKay, Deidre (2005) 'Reading remittance landscapes: female migration and agricultural transition in the Philippines', *Geografisk Tidsskrift, Danish Journal of Geography* 105(1): 89–99.
Potts, Deborah (1995) 'Shall we go home? Increasing urban poverty in African cities and migration processes', *The Geographical Journal*, 161(3): 245–264.
Potts, Deborah (2005) 'Counter-urbanisation on the Zambian Copperbelt? Interpretations and implications', *Urban Studies* 42(4): 583–609.

7 Governing the everyday

Introduction

The discussion, so far, has taken a largely local, agency-focused approach to the everyday. I have been concerned to outline what people do, how they do it, and what this means for livelihoods, identities and processes of social, economic and cultural change. What has been notably missing in much of the discussion is 'politics'. By 'politics', though, I mean politics writ small, or what might be termed 'everyday politics', or 'micro-politics' rather than the high politics which tends to come to mind when we think of governing, government and governmentality.

Kerkvliet (2005) distinguishes three types of politics: official, advocacy and everyday. The first of these is the politics of authorities, whether governmental or linked with other organisations. Advocacy politics involves the challenging, support and criticism of official policies and programmes. Everyday politics, meanwhile, 'features the activities of individuals and small groups as they make a living, raise their families, wrestle with daily problems, and deal with others like themselves who are relatively powerless and with powerful superiors and others' (Kerkvliet 2005: 22). While Kerkvliet considers everyday politics to focus on the 'quiet, mundane, and subtle expressions and acts that indirectly and for the most part privately endorse, modify, or resist prevailing procedures, rules, regulations, or order' (2005: 22),[1] this chapter takes a rather wider view of everyday politics including the ways in which official politics enters and permeates the local and the everyday (and vice versa). What, then, might comprise an everyday politics of the Global South? I see it encompassing three overlapping areas, or arenas.

First of all, there is the question of how far central states and the apparatus of the state can control and influence events in localities. To put it another way, what is the balance of power between state and society? It is significant that even in apparently centralised and bureaucratically heavy states there is often more scope for local level initiative and independence than imagined. This is explored below with reference, particularly, to Vietnam and China. Second, there is a need to consider, in some detail, the role, status and configuration of non-state and semi-state structures at the local level. These may be 'traditional' – such as community-based groups – or 'new' – like NGOs – and they may have local origins or be implants from 'outside'. Whatever their status and origins, however, they represent an important local-level political presence through which national policies are often mediated and they may also act as conduits for local level concerns to be communicated upwards. We should not assume that national policies framed at the centre come to be implemented at the grassroots in the way envisaged and anticipated and the reason for this, often, is because national policies have to be filtered through local gatekeepers. Third, there is the impact of widespread decentralisation on the day-to-day management and administration of affairs at the local level.

As will become clear in the course of this chapter, there is a difficult line to tread when it comes to analysing state–society relations because it is often none too clear where the state ends, and society begins. Having already questioned the utility of other pervasive binaries (see page 57), this chapter does so again with the state/society binary. The state influences social structures, and society can be found imprinted on the state. In Chapter 8 this questioning is taken a stage further to consider the distinction drawn between power/domination and resistance.

In the most part, when it comes to writing about government and governmentality,[2] the emphasis has been to reveal the ways in which the state projects itself into local arenas and to outline, delineate and describe the objectives of that projection (such as development, administration, control, etc.). What is often missing is a consideration of the accomplishments of such a projection of rule. In other words, to ask the question: what does government *do*? The understanding of these accomplishments, Li argues, 'owes as much to the understandings and practices worked out in the contingent and compromised space of cultural intimacy as it does to the imposition of development schemes and related forms of disciplinary power' (Li 1999: 295). In other words: in everyday politics and the politics of place.

The final point to make before embarking on the core discussion is that there is a slightly uncomfortable division between the discussion in this chapter and the concerns of the next which focuses on a critical subset of everyday politics, namely resistance. The intention is that this chapter should provide the explanatory framework for the discussion in the one that follows and, in this regard, the two should be seen as paired.

FROM SOCIETY TO STATE: ACTORS, AGENTS AND ORGANISATIONS

Making people and space legible

When we look at the organisation of societies in the Global South, and particularly their organisation in rural areas, there is a tendency to see the structures and systems as emic and products of long periods of social evolution. It has also been suggested that these traditional structures are sometimes reproduced in urban social contexts. This, to a degree, is the point that Rebecca Elmhirst makes with respect to her interpretation of norms of behaviour and the continuing influence of 'the village' among young, female migrants from rural areas working in and around Jakarta (see page 112). But the street, as ever, is not one way. Writing of Africa, Guyer (1981: 107) argues that the 'state alters the political conditions under which local and kin-based groupings operate, [and] also penetrates them more directly through the legal system'. For example, in Cameroon (see Guyer 1981: 107), Indonesia (Guinness 1994; Silvey 2006) and many other places too, the state has made the nuclear family, from a legal standpoint, the social norm, a tendency which can often be traced back to the colonial period.[3] The effect of this single legal initiative has been, *inter alia*, to undermine customary law, encourage the reallocation of community resources to individuals, alter inheritance patterns, rework tenurial systems, and shift intergenerational and gender relations. Such influences and their effects are unsettling; they could not be anything but. (They may, though, also be broadly positive.) There is the issue, therefore, of considering how far 'traditional' structures at the local level have been affected by outside influences.

In most regions of the Global South, the colonial period saw the introduction of administrative reforms which had the intention of putting people in their place so that they could be counted, mapped, controlled and taxed. Cadastral surveys were undertaken, land was mapped and allocated, spatial units were created (provinces, districts, sub-districts, villages), an administrative structure was mapped onto these units, and there was a progressive territorialisation of people, space and resources (Table 7.1). The 'science' of administration was brought to bear for the first time in many places so that people and resources could be made 'legible' to the state. This has been most fully developed in James Scott's (1998) book *Seeing like a state* in which he explores how scientific forestry, the collection of social data, the mapping and measurement of space, the codification of space (e.g. in terms of land tenure), and the use of surnames have all contributed to making the complexities of the local legible to the state. He argues that:

> These typifications are indispensable to statecraft. State simplifications such as maps, censuses, cadastral lists, and standard units of measurement represent techniques for grasping a large and complex reality. . . . The inventions, elaboration, and deployment of these abstractions represent . . . an enormous leap in state capacity – a move from tribute and indirect rule to taxation and direct rule.
>
> (Scott 1998: 77)

Today, the lowest level of government is usually the village and the lowest level representative of central government is the village head. It is the village head who is normally required to

Table 7.1 Defining terms: territorialisations

Term	*Definition*
Territorialisation	The means and process by which the state extends its control over space, the populations who inhabit that space, and the natural resources found there. People are counted, land is measured and resources are allocated and this is given authority through the 'scientific' approach adopted and the legal structures that underpin the process.
Deterritorialisation	A parallel process to territorialisation by which the state removes local people from the spaces and places they inhabit either in a physical sense (they are resettled elsewhere), or functionally (through the scientific classification of land and its allocation to particular uses), and/or mentally (through endowing land types with particular meanings that override local meanings).
Reterritorialisation	The process by which people insinuate themselves into new spatial contexts, imbuing them with meaning, exerting some degree of control over them, and making them 'home'.
Counter-territorialisation	Attempts by local people to resist the territorialisation tendencies of the state through a variety of grassroots efforts including counter-mapping (in which communities provide their own maps to counter the state's mapping of people, land and resources) and tree ordination (in which trees are sanctified to protect them from cutting). These efforts are often supported and sometimes initiated by NGOs.

Source: Rigg 2005: 109

record and collect vital statistics (births, deaths, marriages), the most basic and elemental of information necessary to inform the legibility project. Of course there were pre-modern forms of authority and leadership, and often what we see today builds on (rather than directly replaces) these antecedents. There was a tendency, for example, for colonial rule to use local practices as building blocks in their edifices of control (Box 7.1). Furthermore, the extension of the writ of the state (in Kerkvliet's terms, 'official politics') into local arenas is ongoing in many areas of the Global South.

BOX 7.1 Ruling Fiji: participatory colonialism in the Pacific

The first governor of the island colony of Fiji, Sir Arthur Gordon, was determined to prevent the dispossession of natives by settlers and, to this end, from 1874 established a system of administration that was based on the traditional Fijian communal system.* A system of indirect rule through village, district and provincial chiefs was introduced with chiefs being actively sought out by the British for their advice and input. Land became the inalienable property of clan groups, and clans-people came under the (now codified) control of chiefs. At one level the British experiment in Fiji was a laudable example of enlightened colonial rule; but it also introduced a system that standardised administration based on a particular reading of 'tradition'.

As Thomas (1994) points out, the traditional model was drawn from those areas of Fiji with which the colonial administrators were most familiar. Chiefs gave advice – which the British listened to and often acted upon – but ordinary Fijians were ignored on the assumption that the chiefs represented the interests of their clans-people and that some chiefs could represent the position of all chiefs. Thomas provides an example of how this codification of custom and normal practice impinged on ordinary people: 'In 1902 . . . because "the Custom of Fiji is to live in villages" the family of a man named Waivure could not live off in a yam garden away from village and clan; their garden house was . . . pulled down' (Thomas 1994: 108). Before village life had become normalised and codified Fijians could be mobile; under the new system, such 'deviations' were to be policed and prevented. Thomas also argues that while the British system of colonial rule in Fiji seemed to preserve traditional structures of power and authority, paradoxically it also increased the control of the colonial state and its ability to govern. It did this in three ways. First of all, by ordering (structuring) native society and communal practice. Through ordering – by which Fijians and Fijian society could become, to the British, knowable – came regulation. Second, by using chiefs as effective and willing channels for control; interlocutors, as it were, between the colonial state and its subjects. And third, by using the communal system as an efficient and effective means of surveillance, channelling valuable information back to the centre.

Source: Thomas 1994: 107–125

Note: * Gordon had been appalled by the effects of colonialism and settlement on the indigenous Maori of New Zealand and was determined that this should not be repeated in Fiji (Thomas 1994: 109).

In the natural resource-rich Lao People's Democratic Republic, the state has introduced policies of land allocation and resettlement partly so that the economic value of forests can be harnessed in the interests of the state and partly so that people can be more easily counted and controlled for reasons of security and development (Vandergeest 2003; Rigg 2005). The policies have the greatest effect in upland areas and among minority, upland peoples. The state has codified the use of space, earmarking some land as preservation forest, some as reserve forest, while allocating other parcels for agricultural activities (see Lemoine 2002: 10–11). Even 'community' forests are designated at the behest of the state rather than being a true reflection of local community autonomy and action (Illustration 7.1). There is now widespread agreement among scholars working on Laos that this process of territorialisation – whatever its intentions may have been – has been destructive and corrosive in terms of livelihoods and, in some cases, poverty-creating. This has been because, at root, state intrusion into upland livelihoods in Laos has unpicked and undermined established livelihood systems:

> The real issue with land allocation and relocation is that the control of individual and communal resources is being wrested away from upland and highland families, who happen to be mostly ethnic minorities. Thus their whole means of livelihood and economic security is being threatened.
>
> (Chamberlain et al. 1995: 42)

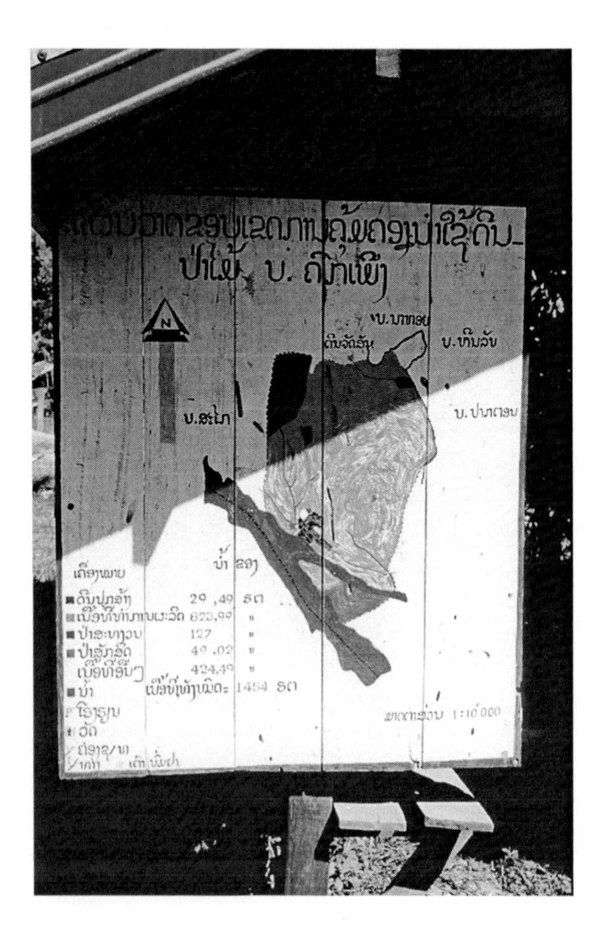

Illustration 7.1 Territorialising land in Laos. Village lands are demarcated, under the orders of the state, according to use

While evidence of local resistance to such territorialisation efforts is scarce in the case of Laos, elsewhere we see people employing everyday forms of resistance to contest and combat state policies and the actions of the state and its minions (see page 171).

Legibility contested and the role of cultural politics

The introduction to this chapter made the point that understanding how state policies and the objectives of rule are worked out in practice owes much to the local setting. Drawing on ethnographic research among 'isolated' people in Kalimantan and Sulawesi (both 'Outer' islands of Indonesia), Li (1999, 2005) shows how state-directed resettlement programmes, in progress since 1950 and designed to bring development and modernisation to remote peoples, have continuously and consistently failed to achieve their stated objectives. Li (1999: 314–315) concludes that to understand not what the Indonesian resettlement programme *aims to do* but rather,what it *accomplishes*, it is necessary to ask a series of questions that are locally contingent: how are the objects of planning framed? How do officials and locals interact and relate to each other? How are 'rules' enforced? What compromises to the dictates of the plan are permitted? And so on. Li suggests that such compromises are normal and regular not because they are expected and therefore planned for, but because the reality of rule involves 'routine and intimate compromises through which relations of domination and subordination are lived' (Li 1999: 316). Li focuses on the local level outcomes of a state legibility project – namely land settlement in Indonesia – and having found it to be compromised at every turn raises questions about the ability of the state to accomplish its stated aims.

In another study from Indonesia, Bebbington et al. (2004) reached much the same conclusion. Drawing on fieldwork in two villages in Central Java and one in Jambi province in Sumatra, they show how village politics in Indonesia is tightly bound up in local religious tensions in the first two villages where orthodox Islam is in competition with *abangan* forms of the religion,[4] and local land disputes in the Jambi case where *adat* (local custom) is being used to reintroduce traditional systems of governance in the face of a wave of land-hungry migrants. Intuitively one imagines that it is hard, perhaps impossible, to isolate local governance as a technical and administrative process from the cultural context within which it is nested. Governance, for Bebbington et al. (2004: 189), 'is a cultural as much as an administrative process . . . [and] it is also important to understand that the local is a site of struggles over cultural practice'.

Both Li (1999) and Bebbington et al. (2004), then, are concerned with understanding how politics *operates* in localities, seeking to understand what happens and why. They are both working at the interface between politics and culture, or in the field of 'cultural politics' which involves the exploration of what is cultural about politics, and political about culture (Armitage et al. 2005: 1). Rather than seeing local government and administration as a technocratic and managerial 'business', there is a case for their deeper contextualisation in the local cultural setting. If this happens then the failure (or success) of initiatives is not reduced to technicalities such as lack of capacity but seen in terms of the ways in which the formal institutions of local government sit within, affect, and are affected by 'culture'. It also raises the likelihood that whether we choose to look at local level governance from the perspective of the state or from the vantage point of society, will result in very different views indeed.

NGOs, POs, GROs, CBOs, VOs and other acronyms[5]

Along with a reignited interest in traditional and local structures and their roles in local governance has come another revolution: the NGO revolution. It is not possible to put an exact figure on the number of NGOs around the world except to note their phenomenal growth and proliferation (Table 7.2). To this explosion in NGOs, can be added the growth in an assortment of other types of local organisations, creating an alphabet soup of groups operating at the local level.[6] NGOs are important not only in terms of their contribution to 'development' – which is how they tend to be regarded in the North – but also, and perhaps more importantly, because of their political role: 'NGOs have become inextricably implicated in civil society, democracy, good governance and social capital' (Desai 2002: 497). Indeed, the tendency to ignore NGOs as a political force and instead concentrate on their development remit can be regarded as problematic.

While many NGOs may explicitly focus on development issues in the Global South, the means by which they do this is often intensely political, albeit usually in a low visibility way. Their emphasis on participation, empowerment, and alternatives to existing (state-led) methods and structures places their actions in the field of politics. Moreover, because many NGOs operate at the local level, whether in rural villages or urban settlements and with peasants or workers, places them four-square in the realm of everyday politics. At the same time, however, NGOs are often linked into national and global systems. These systems embrace both issues of financing (significant amounts of overseas development assistance is channelled through NGOs, see Table 7.2) and the ideas and ideologies which underpin much NGO activity. For some scholars and activists, NGOs have been mainstreamed, tamed and domesticated so that today they act as conduits for the ideas, agendas and initiatives of the neo-liberal international institutions. For others, there is still space for alternatives (Townsend et al. 2004). Wherever one stands, however, there is a 'trans-national' community of non-governmental organisations (Townsend 1999), so that NGOs are intentionally local and also, often, importantly trans-local.

In the department (or region) of Cusco in the Peruvian Andes, some 68 NGOs channelled between US$90 and US$100 million into local projects during the 1990s. This compares with

Table 7.2 The NGO revolution

Date	*Number of NGOs*	*Assistance channelled through NGOs*
1980	608 NGOs in consultative status with the UN	
1988		US$4.2 billion channelled to South
1989	4,000 NGOs in OECD countries	US$6.4 billion channelled to South
1990		US$7 billion of international development funds channelled through NGOs
1993	28,900 international NGOs	
2000	37,281 international NGOs; 2 million NGOs in India alone	
2002	2,236 NGOs in consultative status with the UN	

Note: data on NGOs are notoriously unreliable and inconsistent. The figures in this table are provided to give a rough indication of scale and growth over time.

Sources: Desai 2002 and other sources

bilateral and multilateral assistance totalling around US$35 million (Bebbington 2005: 940). Much of these NGO funds came from Catholic funding bodies in Europe and, particularly, from the Dutch co-financing agency Cambio. Bebbington regards 'Dutch foreign assistance [having] a profound effect on the structure and nature of development organizations in Cusco' (Bebbington 2005: 940–941). There is an 'aid chain' which links NGOs operating in local areas of Cusco with funding pressures and demands in the Netherlands. One outcome, it has been suggested, has been a deradicalisation of development in Cusco and, arguably, a depoliticisation of development. But if we consider the political in the everyday terms which is the focus of this chapter, there opens up a space for radicalism even within a normalising, managerial context.

In a series of publications, Janet Townsend, Gina Porter and Emma Mawdsley (Townsend 1999; Mawdsley et al. 2002; Townsend et al. 2004) have provided a comparative exploration of the roles of NGOs in Ghana, India and Mexico. They note how hard it is to 'read-off' whether an NGO will be progressive or reactionary in terms of its effects. This is because of the disjuncture between what a NGO appears to be as an organisation, and the 'submerged networks in the democracy of everyday life' which makes it both less and more than it appears (Townsend et al. 2004: 885). This means that reactionary, patriarchal NGOs may be – in practice – liberating for some women in some places, while progressive NGOs may fail to engender the empowerment that they plan for. In part, this is for the reasons that Bebbington et al. (2004) and Li (1999, 2005) highlight in their work on Indonesia noted earlier: the everyday has a habit of getting in the way, so what NGOs and governments *aim* to achieve is often very different from what they actually *accomplish*.

Problematising local participation

One of the key areas where NGOs have tried to influence the day-to-day working of local politics is through widening and deepening 'participation'. Participation is defined as a 'process through which stakeholders influence and share control over development initiatives, decisions and resources which affect them' (http://www.worldbank.org/participation/participation/parthistory.pdf: 1; and see Francis 2001). From being radical and subversive in the 1950s when social workers and activists, particularly in Latin America, bemoaned the absence of broad-based community involvement or participation in their own affairs, by the 1980s participation had become part of mainstream orthodoxy and a panacea for many of the ills seen to be facing the Global South (see Cleaver 2001: 36):

> The euphoric word 'participation' has become part of development jargon. No respectable project can not use this 'in' word now, nor can it get funding without some provision for the participation of the people.
>
> (White et al. 1994, quoted in Michener 1998: 2105)

By the early 1990s, the 'mainstreaming' of participation was such that the World Bank had set up a 'learning group on participatory development' (in December 1990) and it had become central to the operation of most multilateral and bilateral organisations working with ordinary people in the Global South.[7] Participation does not begin and end with development, however. Participatory democracy involves opening up arenas of governance so that ordinary people can become more fully involved in the decisions that shape their lives (Cornwall 2002). This may mean opening up existing and official political spaces to popular participation, boosting citizen engagement and people's governance. But, as Cornwall

(2003: 51) notes, such 'officialised' spaces are matched by unofficial spaces and the spaces of the everyday. These unofficial and everyday political spaces are more fluid and contingent than official spaces because they are continually created and reformulated over time. While from a local perspective they may be structured and meaningful outcomes of local needs, from an official standpoint they often seem chaotic, even anarchic.

Inevitably perhaps, there has been something of a backlash against the uncritical embracing of participation. Scholars began to see that participation, in practice, sometimes led to outcomes which were not particularly participatory (Table 7.3).[8] What concerns me here, however, is one particular area of this wider critique of participation because it relates directly to the governance theme of the chapter and, more particularly, the articulation of 'new' and 'old' political structures, namely, 'participatory exclusions'.

India's Joint Forest Management programme and other participatory exclusions

Forests and community forests or commons have long played an important role in livelihoods, supplying fuel wood, fodder and various non-timber products. They seem to have been particularly important for poorer sections of rural society and for women. In India and Nepal, the decline in these forest resources reached 'crisis proportions' by the early 1970s, ushering in various social forestry programmes (Agarwal 2001: 1625). Many of the state-directed efforts, based on exotic species such as eucalyptus and implemented in a top-down fashion,

Table 7.3 Levels of modes of participation

Form/level of participation	*Characteristic features*
Level I: nominal participation	**Legitimation**
Nominal participation	Membership in a group
Passive participation	Being informed of decisions *ex post facto*; or attending meetings and listening in on decision-making, without speaking up
Functional participation	Enlisting the support of local people to minimise dissent and maximise compliance
Level II: instrumental participation	**Managerialism**
Consultative participation	Being asked an opinion in specific matters without guarantee of influencing decisions
Instrumental participation	Enlisting 'volunteers' to take part in tasks and activities; delegating responsibility
Active participation	Expressing opinions, whether or not solicited, or taking initiatives of other sorts
Level III: transformative participation	**Empowering**
Interactive (empowering) participation	Having voice and influence in the group's decisions
Transformative participation	Building political capabilities, critical consciousness and confidence; demanding rights; enhancing accountability

Sources: collated and combined from Agarwal 2001: 1624 and Cornwall 2003: 1327

failed. By contrast, community-based initiatives drawing on indigenous knowledge, local species and community action were viewed as successful. From this arose a consensus that 'successful forest management needed the participation of local communities' often in the form of Community Forest Groups or CFGs (Agarwal 2001: 1625; see also Corbridge and Jewitt 1997). In India, this consensus has been reproduced through the Joint Forest Management (JFM) programme introduced from 1990. Essentially, the JFM brings together the government, NGOs (sometimes, as intermediaries) and local people in the regeneration of degraded forest and permits each to share responsibility and reward. By the end of 2003 the number of JFM groups in India was 84,632, managing 17.3 million hectares of forest out of a total official forest cover of 67.5 million hectares (www.rupfor.org/jfm_national scenario.asp).

Joint Forest Management in India is presented in much of the literature as a success story in participatory forest management and rural development. But while the programme may go under the umbrella term of 'community participation' it represents a formalised system which has often replaced other and older forms of communal management. To see whether the rhetoric of community participation is reflected in the reality of forest management, Agarwal (2001: 1626) poses the central question: 'Are these systems inclusive and equitable in relation say to women, especially the poor?'.

JFM is structured at the local level around two bodies: a General Body (GB) which can, in theory, draw its membership from the whole village and an Executive Committee (EC) of between 9 and 15 people. These two bodies determine rules of local forest use and therefore the make up of their membership is crucial. Agarwal found that women comprise just 10 per cent of JFM general bodies. Usually this is because of the way they are constituted: often only one member is permitted per household, and that one member tends to be the male head of household. In the Executive Committees, women's membership is also low – in the 20 JFM groups studied in West Bengal by Agarwal, 60 per cent of their ECs had no women members. Moreover, landless households had almost zero representation. Finally, those few women who were counted among the membership rarely contributed to discussions. JFM General Bodies and Executive Committees became, de facto, 'men's groups', even when there were women present (Agarwal 2000: 286). Agarwal calls these tendencies 'participatory exclusions' – 'that is exclusions within seemingly participatory institutions' (2001: 1623). The outcome is that women's specialist knowledge of the forest is overlooked, inequitable distribution norms and rules tend to be reproduced, and forest closure policies tend to be conservative, overlooking the needs of women. So, for example, regarding the last of these, more than half of the 87 CFGs that Agarwal studied in 1998–1999 had banned the collection of firewood, more than doubling collection time for a headload of wood to 4–5 hours in some areas and forcing women to steal wood from protected forest. 'We don't know in the morning how we will cook at night', poor, low caste women in Uttar Pradesh told Agarwal (2000: 286). The problem, of course, is that the JFM programme does not create any new mechanisms for including women in the process of forest management. Even when efforts are made to include women, these may, in reality, change little because they do not lead to any change in the structures that exclude *most* women even if *some* women find themselves on the necessary committees.

Agarwal's concerns and assertions are well founded and powerful, but they are also built on certain taken-for-granted assumptions about the nature of environmental knowledge and, more particularly, gendered environmental knowledge. Corbridge and Jewitt (1997), working in the village of Ambatoli in Bihar, reveal that only a minority of households knew about traditional forest management systems and women's knowledge was rarely passed from mother to daughter. In their view, environmental knowledge was partial and the 'villagers who were best informed about the jungle were men' (Corbridge and Jewitt 1997: 2158). This was particularly true for the Backward and Scheduled caste households (i.e. the poorest) among whom it is the men – rather than the women – who collect fuelwood.

Agarwal (2000) sees participatory exclusions, and particularly those that relate to gender, being reproduced in many countries and contexts and makes a case for their general applicability. In the case of water management in Thailand, Resurreccion et al. (2004) interpret affairs in a similar manner:

> We argue that participatory approaches . . . may tend to overlook and conceal power relations in general, and unequal gender relations in particular. Thus current participatory practices within regional environmental governance, while basically well intentioned, may inadvertently build on and reinforce social inequalities.
>
> (Resurreccion et al. 2004: 522)

Much like Agarwal (2000), Resurrecion et al. (2004) come to this conclusion because of the way in which water management in northern Thailand is a reflection of wider discourses about gender norms. Women may de facto farm, but discursively they are referred to as 'housewives' or, possibly, 'traders'. Water management, therefore, becomes men's business and women are excluded from membership of the local-level *müang fai* irrigation committee (IC). The authors conclude:

> Women's participation at the community level is influenced by the gender division of labour and discourses on women's 'place' in society. Their participation is seen as an extension of their domestic roles and responsibilities.
>
> (Resurreccion et al. 2004: 526)

For radicals, participation has been emasculated because it has been embraced by governments and mainstream institutions (like the World Bank) and become, in the process, a technocratic and apolitical means of achieving community management and development. Chhotray (2004), in her study of a participatory watershed development project in Andhra Pradesh (India), writes: 'In KWO's [Kurnool District Watershed Office] scheme of things, participation in the project is structured as an itemized protocol akin to the physical and financial targets of the action plan' (Chhotray 2004: 343). KWO does not so much ignore politics, however, as re-cast it as 'politics-without-conflict'. The 'unanimous' agreement of the 'community' is security through 'public' meetings where 'participation' occurs between community members of equal status and standing (Chhotray 2004). Participation becomes a rubber stamping exercise to lend legitimacy to state-directed initiatives.

There is, however, another way of looking at participation and participatory exclusions. That is to see participatory approaches operating in thrall to local political cultures. If we take this view then politics is not so much negated as re-cast in local terms. What we see in these examples from India and Thailand, and in the last section from Indonesia (and see Box 7.2 for a case study from Ethiopia and Mozambique), is the way in which three levels or types of political activity intersect. First, there is the official politics of the state and the organisations of the state and the ways in which they infiltrate local areas. Second, there is the 'new' politics of local participation. Third, there is the 'old' cultural politics of the locality. Behind these three types of political activity are various bodies and groups, which are constituted in various ways, and which garner their legitimacy to speak for local people using different

BOX 7.2 Natural resource management in Ethiopia and Mozambique

As in India, natural resource management efforts in Africa have increasingly turned to indigenous, community-based structures and institutions on the grounds that these are community based, have cultural and historical legitimacy, and will represent the interests of 'the people'. Working in Manica Province of Mozambique and Borana District in Ethiopia, Black and Watson (2006) question this perspective and the efforts that have arisen from it. Their critique is centred on four problematic assumptions about such institutions, each of which they question:

- the assumption that such grassroots institutions are, indeed, indigenous and traditional (almost, timeless)
- the assumption that they are territorial – i.e. that they have power and authority over the management of space
- the assumption that they are apolitical and operate 'outside the world of politics'
- the assumption that they are representative of 'the people' and their interests.

These misplaced assumptions lead to problematic outcomes. First, they lead to the exclusion of some people and groups from access to natural resources and this, in turn, can lead to conflict and violence. And second, the denial of the political nature of such institutions will lead them, in their own terms, to fail.

Source: Black and Watson 2006

means. Politics at the local level occurs at the nexus of these political forms and groups and while in theory it is easy enough to separate them out and treat them individually, in practice they form a political mélange.

An example comes from Williams' (1997) work on state–society relations in contemporary West Bengal, drawing on fieldwork undertaken in 1992–1993 in three villages in Birbhum District. Williams is concerned to challenge the assumption, prevalent in India and in scholarship on India, that there is a sharp divide between an elite-dominated developmental Indian state on the one hand, and 'the masses' on the other. He does this through an examination of the effects of *Panchayati Raj* (local government reform) at the local level. The inhabitants of the three study villages were generally sceptical about the ability of the reforms to increase popular participation or radically to change the nature of politics at the local level. Higher level patronage networks were being partially reproduced at lower levels. That said, the very fact that sometimes poor villagers were being elected into positions of power as local government representatives, and the additional fact that they were required to act as traditional as well as modern leaders was 'blurring the distinctions between the actions of the "modern" state and the negotiations of "traditional" village politics' (Williams 1997: 2107). At the grassroots, then, local government reforms were not so much modernising village-level politics through the state's penetration of society but instead bringing together 'upper' and 'lower' political discourses (those of state and society) in a hybrid political amalgam where the state and the everyday intersected and merged.

A final and even more striking example comes from McEwan's (2005; see also McEwan 2003) study of post-apartheid citizenship in South Africa. Working with civil society organisations and local people in rural and peri-urban settlements in the Western Cape and KwaZulu-Natal, McEwan was interested in exploring how constitutional rights to equality are enacted at the community level. She found that community forums and councils are usually founded on local cultural norms and will reproduce the participatory exclusions of traditional society. Her Xhosa women respondents in the Western Cape were excluded from membership of local organisations and discouraged from participating and speaking publicly (as were the young, so there was a generational as well as a gendered pattern to exclusion). 'In South Africa', she writes, 'the ways in which newly created structures connect with existing institutions, either "traditional" governance structures or local associations, are significant in reproducing existing relations of exclusions that further marginalize groups such as women and young people' (McEwan 2005: 976). She quotes one of her respondents saying: 'Women are expected to keep quiet in meetings. We end up with football pitches instead of crèches' (McEwan 2005: 976). New spaces of citizenship do exist in South Africa, but in the most part only bureaucratically and *de jure*. Practically and de facto their operation is compromised by cultural politics.[9] Not until *D*emocratic transformations at the top are matched by *d*emocratic transformations at the bottom will the opportunity for democratic participation be matched by spaces of everyday participatory democracy, particularly for the poor, excluded and marginalised. It is for this reason that radical feminists challenge approaches to empowerment that blame the victim by assuming that women's lack of power is due to their lack of education, experience, confidence, and money-making opportunities: 'it is in the nature of empowerment that it cannot be given. It has to be taken. If we [African women] wait for male patriarchal government to give power to women, we shall wait for ever' (Longwe 2000: 30).

What can we say in summary about these various case studies drawn from work in Africa and South and South-East Asia? Essentially, three things. First, it is rare to find local level systems of governance that are not shot through with, and implicated in, non-local agendas. Second, local level discourses (for example, concerning the respective roles of women and men) disturb and distort governmental (official) and non-governmental management processes and practices. Third, both these things mean that the existence of two assumed binaries – which often provide a template for analysis – are questionable: the local/non-local binary of scale; and the traditional/official binary of power and control.

Giving and taking: decentralisation fever

The 1990s saw the globe catch decentralisation 'fever' (Bebbington et al. 2004). Decentralisation involves the transfer of power, authority, responsibility, budget and personnel from central government agencies to lower tiers of government.[10] This was spread by an assortment of assumptions and desires. The assumption that government should be decentralised as far as possible and reasonable for efficiency's sake; the belief that development is best achieved if it is controlled by local people (the 'subjects' of development); and the wish to increase popular participation and democratise the business of government and administration. Embedded in these desires are two core justifications: a utilitarian or managerial justification that local level government is better than higher levels of administration in delivering services and amenities (schools, roads, health facilities, etc.); and a political or ideological justification that government *should* be decentralised and put in the hands of local people as part of a process of democratisation. The first of these has been called the 'pragmatic' school of decentralisation (Schönwälder 1997: 757–759) because it concentrates on how decentralisation can promote local and regional development. What the pragmatic

approach also does, however, is depoliticise decentralisation by seeing administrative reforms as largely technical. Critics point out that decentralisation is 'by definition' political because it involves the rescaling of power and 'all of the various possible outcomes of decentralization are also inherently political' (Schönwälder 1997: 758). These critics, instead, place politics at the centre of their approach to the analysis of decentralisation.

Since the highpoint of decentralisation in the mid-1990s it has increasingly come to be recognised that there is no a priori reason why decentralisation should achieve these aims or do so any better and more effectively (or, even, democratically) than higher levels of government (Schönwälder 1997; García-Guadilla 2002; Bebbington et al. 2004). Venal, rapacious and unrepresentative bodies may exist at local levels too. Indeed, decentralisation and democratisation do not go hand-in-hand and decentralisation – like participation – tends to become caught up in local-level political tensions and asymmetries. Decentralisation often seems to result in resources and information being de facto privatised and this then creates increased opportunities at the local level for the powerful to benefit and thrive, often at the expense of the poor and marginal. Decentralisation, therefore, when there are deep-seated inequalities in terms of wealth and power, and when local level institutions are dominated by elites will only further empower those elites while permitting them to capture an even greater share of resources.

Like many countries in the Global South (and North), the government of India has introduced far-reaching policies of decentralisation and devolution. This process began in 1993 with a series of constitutional reforms that devolved power and authority to district, sub-district and village level bodies (Johnson et al. 2005: 939). Drawing on fieldwork undertaken in 2001–2002 in six villages and three districts in each of the two Indian states of Andhra Pradesh and Madhya Pradesh, Johnson et al. (2005) show how the outcome of this devolution process in terms of local governance and, more particularly, the advancement of poor and marginalised groups, is unpredictable.

In Madhya Pradesh, between 1994 and 2003, reforms have increased the power of the *gram sabha* (village assembly) and the *gram panchayat* (village council). In 2001, the state introduced *gram swaraj* or 'village self-rule'. By contrast, in Andhra Pradesh, democratic institutions at the local level have been by-passed by the state government in favour of non-elected bureaucrats. Johnson et al. (2005) then examine the effects of these different approaches for politically marginal groups at the local level. They come to two main conclusions. First, that the devolution of authority to local bodies is not sufficient, in itself, to empower marginal groups. And second, that the direct engagement of marginal groups with non-elected representatives of the state may offer a productive and effective means for the largesse of the state to be channelled to the poor and excluded. What the work shows – and this resonates with much else in this chapter – is that creating the political structures of decentralisation and devolution is not sufficient, nor even necessary, for creating empowering and equitable local government.

The discussion over decentralisation often begins with the premise that the motivation – in policy terms – is the laudable desire to empower local groups, more effectively deliver amenities and services to local people, and to do this in a fashion that accords with local wishes and desires. There may be other, less virtuous motives and, indeed, other, unintended and less desirable outcomes. Thomas and Twyman (2005) examine the decentralisation of rural water supply in Namibia. In line with the paradigm of community-based natural resource management noted above, the Namibian Directorate of Rural Water Supply aims to have all rural water supply under community management by 2007. This, however, is primarily a cost-cutting exercise: the costs of provision and management are passed from the state to local Water Point Committees. Further (and this is the unintended consequence), in some areas the effect has been to marginalise poorer people. Unable to meet their share of the costs of maintenance and upkeep, the poor either have to donate their labour or they may find access denied.[11] So, decentralisation in this case can lead to new participatory exclusions emerging as well as having a negative effect on equity.

State–society relations in Vietnam

The discussion so far in this chapter has focused, essentially, on two areas: first, how far the state can and does influence society; and second, how far and with what effects decentralisation and participation has reconfigured local level politics and governance. There is also a third and rarely commented on possibility: that society influences the state.

For some time there has been a discussion of the relationship between state and society in Vietnam and this has become particularly pertinent with the economic reforms (*doi moi*) introduced from 1986. Essentially, scholars have taken one of three positions (Kerkvliet 1995b: 398–399) on the issue:

- That the Vietnamese state is powerful, dominant and domineering and there is 'little scope for the organisation of activity independent of the party-led command structures' (Thayer 1992: 111, quoted in Kerkvliet 1995b: 399).
- That the Vietnamese state does respond to social pressures from below but only through organisations of the state at the local level.
- That society and social forces outside the immediate control of the state are much more influential than imagined and that the state not only has to navigate its way through these local-level structures but also responds to them.

Based on four periods of fieldwork undertaken between 1992 and 2000 in rural sub-districts south of Hanoi and drawing on a wealth of archival material, Kerkvliet (1995b, 2005) makes a sustained and convincing case for embracing the third of these positions. It is worth quoting from the conclusion to his book at some length:

> Everyday politics matters. That, in a nutshell, is a central conclusion to draw from this book. It can have a huge impact on national policy. Consider what happened in Vietnam. Collective farming, a major program of the Communist Party government, collapsed without social upheaval, without violence, without a change in government, without even organized opposition. Yet national authorities were pressured into giving up on collective farming and allowing family farming instead. To a significant degree, that pressure came from everyday practices of villagers in the Red River delta and other parts of northern Vietnam. To a significant extent, those practices were political because they involved the distribution and control of vital resources. Those everyday political practices were often at odds with what collective farming required, what authorities wanted, and what national policy prescribed.
>
> (Kerkvliet 2005: 234)

What is surprising – and on first glance remarkable – is that an apparently centralised, bureaucratic polity, a one-party communist state, was pressured from below into rejecting one of the ideological and economic pillars of national policy, namely collective farming. It was not just that the writ of central government failed to penetrate the local level; the actions of peasants were such that their views and opinions resonated all the way up and back

to the leadership in Hanoi, ultimately forcing a change in policy. This raises specific questions about whether Vietnam can be regarded as a 'strong' or a 'weak' state (see Painter 2003, 2005). More interestingly, though, it raises questions about whether the strong/weak state binary is very helpful when it comes to interpreting politics and governance at the local and everyday levels. Rather than being representatives of central government in rural areas, cadres in the North Vietnamese countryside were – and are – local people with local affinities and local sensitivities. Cadres often, it seems, turn a blind eye to 'fence-breaking' by the masses as they challenge the writ of the state while the leadership in Hanoi are highly sensitive to local level demands and protests (Painter 2005: 266–267). Moreover, if we combine this with some of the points made earlier with regard to local level cultural politics (in Indonesia, India and South Africa) then we arrive at a position where the room for local manoeuvre is much greater than one would imagine in a state which, far from being centralised and bureaucratic is, in practice, characterised by a 'highly decentralised, fragmented and sometimes incoherent set of state institutions' (Painter 2005: 267).

The insidious role of the state: where is the state in the everyday?

From the evidence and examples presented so far, it would be tempting to conclude that the local and the everyday play a much more significant role in politics and government than state-centric views tend to permit. It has even been suggested that in Vietnam there is a strong case that local political pressure forced a centralised, one party state to rescind and reverse a central plank of policy. But to leave the discussion here would be to over-privilege and reify the power of the local. To remedy this situation, the discussion will briefly turn to consider how the state infiltrates the grassroots, and how local institutions and structures can be domesticated and bureaucratised by the state. To achieve this, however, rather than turning attention to high-level politics as a counterpoint to the earlier discussion, the focus will be on examining how – and how far – the state and state policies infiltrate local spaces and determine local livelihood outcomes.

In Chapter 6, Box 6.1 (page 123) outlined the way in which the *hukou* system in China segments the Chinese population into rural and urban. There are two particular issues of this system which can be brought to bear in the context of the discussion in this chapter. First, mobility and labour market segmentation (sectoral and spatial) can be seen to be governed by the gatekeeping role of the state (Fan 2001, 2002). Second, this is clearly reflected in the livelihood outcomes of migration in China. In her work on Guangzhou, Fan (2001) surveyed (in 1998) 305 non-migrants (i.e. urban residents), 300 permanent migrants, and 911 temporary migrants. Those who had been awarded the coveted status of 'permanent migrants' from rural areas because of their skills, education or links with the state had the greatest incomes, the best benefits, and the highest education (Table 7.4). Fan concludes that 'most importantly, the empirical analysis has shown that resident status [bestowed by the state] exerts compelling effects on labour-market returns even after effects of achieved attributes such as education are held constant' (Fan 2001: 504). Temporary migrants are not only effectively excluded from better-paying jobs by the *hukou* system but also denied many of the benefits (health care, retirement benefits) that come with these jobs. The state, in this way, has created an urban labour market that is highly segmented and highly biased in favour of those with urban resident status: 'resident status operates much like ascribed attributes such as race and sex [gender] in Western labor markets – a status that it is very difficult to change but exerts independent effects on labor-market returns' (Fan 2001: 505).

Table 7.4 Income, benefits and education by migration status, Guangzhou, China, 1998

	Urban non-migrants	*Permanent migrants*	*Temporary migrants*
Urban place of birth (%)	91	69	30
Professional occupation (%)	17	46	7
Mean monthly income (yuan)	1,836	3,654	1,511
Medical benefits (%)	60	71	8
Retirement benefits (%)	49	63	4
Free lodging (%)	1	17	60
Senior high or above education (%)	29	78	10
Illiterate or primary level education	9	3	17

Sources: extracted and collated from Fan 2001, 2002

In contrast to Vietnam where peasants managed – at least on one reading of the evidence – to challenge the state, rework policies to their benefit and, ultimately, to force a reversal of those policies, in China the *hukou* system continues to play a far-reaching and insidious role in determining livelihood outcomes at the local level. By segmenting and then cementing the population into rural and urban classes the *hukou* system prevents spatial, occupational, economic and class mobility. At the extreme it could be said to consign some people to poverty. Rural migrants come to be considered as 'outsiders to the urban society, and most are relegated to the bottom rungs, picking up dirty, dangerous, and low paying jobs and finding a marginalized and underclass existence in the city' (Fan 2002: 107).

But even in China, not all policies are effectively and seamlessly transmitted to local levels. Like Vietnam, this assumes that townships will implement the policies of the centre and that, in turn, village cadres will respond in the ways envisaged by those policies. Mood (2005) explains the strikingly different development paths that villages in China have taken not (just) in terms of resource endowments, historical legacies, geographical location, and so forth, but in terms of the way in which differences in village–local state relations shape different paths.

The *hukou* system stands out as a particularly clear example of the gatekeeping role of the state. We also see the state involved in a more diffuse manner, connected with the production of certain gendered labour processes and migration streams. Taking a state-centric view of 'women's work', and it is possible to argue that gendered labour processes are a product of the operation of the capitalist, patriarchal state as 'women' are reconstituted as 'workers'. Adopting a society-centric position, however, merely serves to highlight the limitations of state power. This everlasting dance between a disciplinary state and a variously resistant society, and the difficulty (arguably, impossibility) of arriving at a 'correct' vantage point is exemplified in the case of female domestic workers in Indonesia where the state has created a bureaucratic and cultural space for such work to operate and also for such work to be challenged (Box 7.3). It is also reflected in the way in which the majority of the inhabitants of the city of Karachi in Pakistan live and work (Hasan 2002). Most Karachiites live in informal and illegal settlements occupying public land. These settlements are developed by private land developers who operate with the connivance of government officials, and are protected from police action through bribes. The developers are also, however, instrumental in the creation of residents' associations to lobby government for services and amenities and for the regularisation of tenure for 'their' illegal settlements and the residents of these settlements. They go so far as to hire journalists to write stories about the poor conditions as a means of exerting pressure on the authorities. It is not just housing which operates in the everyday and in-between spaces of formal and informal, state and society, and legal and extra-legal. It extends in the case of Karachi to schooling, travel, health facilities and employment. Indeed, to almost all the spaces of the everyday (Figure 7.1).

BOX 7.3 Domesticity and domestic workers

The Indonesian New Order (1966–1998) state under the leadership of President Suharto fashioned and moulded women into filling two roles: the role of mother and home-keeper in line with state ideologies of feminine domesticity; and the role of low income worker to achieve the state's development project. These two roles do not sit easily. The role and image of the ideal woman as home-maker, mother and wife is centred spatially and functionally on the home. That of the female worker (nimble-fingered, hard-working and controllable) is centred on the factory and the place of work. The state got around this tension, arguably contradiction, between women as workers and women as home-makers in two ways. First, by injecting a class division into the equation: home-makers would be middle class, nuclear families; workers would be peasants and, over time, become working class. Second, by suggesting that women could play the dual role of home-makers and workers but only so long as work did not interfere with their domestic responsibilities. This often meant a generational shift from women as workers, which operated from adolescence to motherhood, to women as home-makers, which came into play from motherhood onwards.

With this worker/home-maker tension rather uncomfortably reconciled, from 1983 the Indonesian government began to permit and encourage Indonesian women to travel to Saudi Arabia to work as domestic servants. In the five years 1990–1994 almost 400,000 Indonesian women officially took up work in Saudi Arabia with a target figure of 1.25 million for the five years 1994–1999. (The true figure is likely to be much larger given the number of workers who travel to Saudi Arabia without formal documentation.) Almost from the day that the migrant stream began to flow, reports of maltreatment, physical and mental abuse, and exploitation started to filter back to Java.* In 1984 the first NGOs began to campaign on behalf of these women migrant workers. The women themselves would also actively struggle against their conditions, breaking their contracts and even assaulting their employers in self-defence. One of the reasons why the reports circulating in the Indonesian national press received such a willing and supportive readership was because of the New Order state's very own construction of idealised female domesticity. This culminated in March 2005 with the

continued

suspension by Indonesia of the sending of unskilled labour to Saudi Arabia. The process resumed only in August 2005 when a bilateral agreement was concluded between the two countries setting out minimum standards of work, conditions and pay.

What we see in this case study is an insight into both the extent and the insidious nature of state power and, at the same time, the limits of that power. The Indonesian state has managed, through its development and education programmes, to recast women as wives/mothers and workers. Through their relations with the government of Saudi Arabia, the New Order government created a new gendered migration stream of low wage, female domestic workers. But the abuses these women faced in turn resulted in a backlash that was orchestrated by NGOs which used the New Order's own idealised image of feminine domesticity and its claimed paternalism to protect these workers from abuse. As Silvey writes: 'state power relies on deployments of particular gendered spaces, subjects, and scales, but it is through these constructions that it can be challenged' (Silvey 2004: 261).

Principal source: Silvey 2004

Note: * To read some of these personal stories of physical and sexual abuse and harassment, including torture, see the websites of organisations such as Human Rights Watch (www.hrw.org), Amnesty International (www.amnesty.org) and the International Labour Organization (www.ilo.org).

From indigenous knowledge through social capital to community development: a new prescription for government?

This chapter has suggested that there are a series of links – sometimes tenuous – between indigenous knowledge and community structures through to modern forms of governmentality as they operate at the local level whether these are state orchestrated or constructed and mediated by NGOs or other community level groups.[12] This sequence of loose associations is presented diagrammatically in Figure 7.2.

Local knowledge, structures, cultures and populations are increasingly used by states to govern, and by NGOs to manage. While in the past local knowledge and culture was ignored and sometimes actively overwritten (not least in achieving 'development'), today governments, international agencies and non-governmental organisations take note of, and pay homage to, culture (Radcliffe and Laurie 2006; see also Radcliffe 2006). Culture – notwithstanding the definitional difficulty of pinning the term down – has become a resource to be accessed and exploited not only in the achievement of material development objectives, but also in the wider fields of government and planning.

There are three aspects of this tendency which the chapter has sought to highlight. First of all, there is the way in which particular views and interpretations of local culture are gathered, simplified and then codified so that they can become part of the architecture of control and governance. The fact – like Joint Forest Management in India or colonial rule in Fiji – that these systems can be presented as 'local', 'indigenous' or 'community-based' gives them a stamp of legitimacy. But this localism may not be warranted and therefore the legitimacy may be, in part, spurious or, at least, arising from elsewhere. For this reason, the meaning(s) of culture is/are contested. Second, there are the gaps that exist in systems of administration,

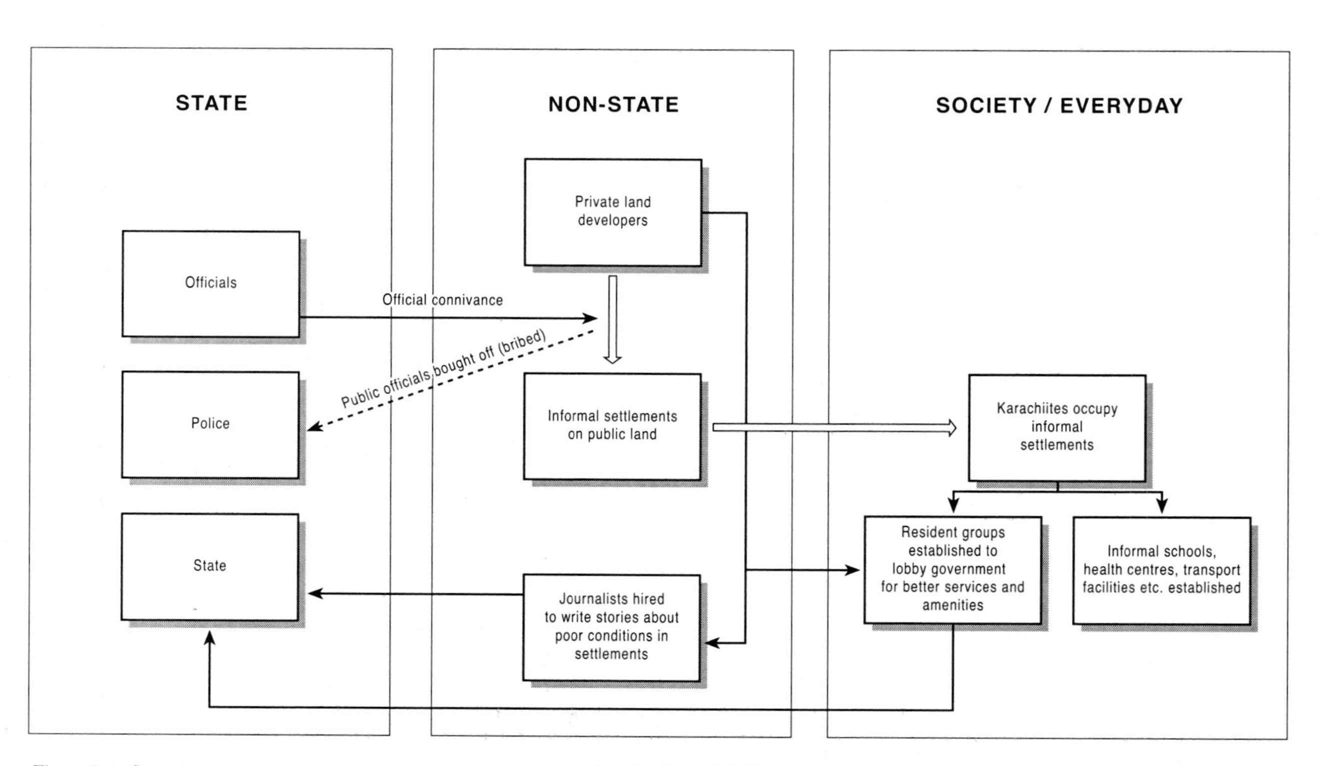

Figure 7.1 Crossing boundaries: state, society and the everyday in Karachi, Pakistan

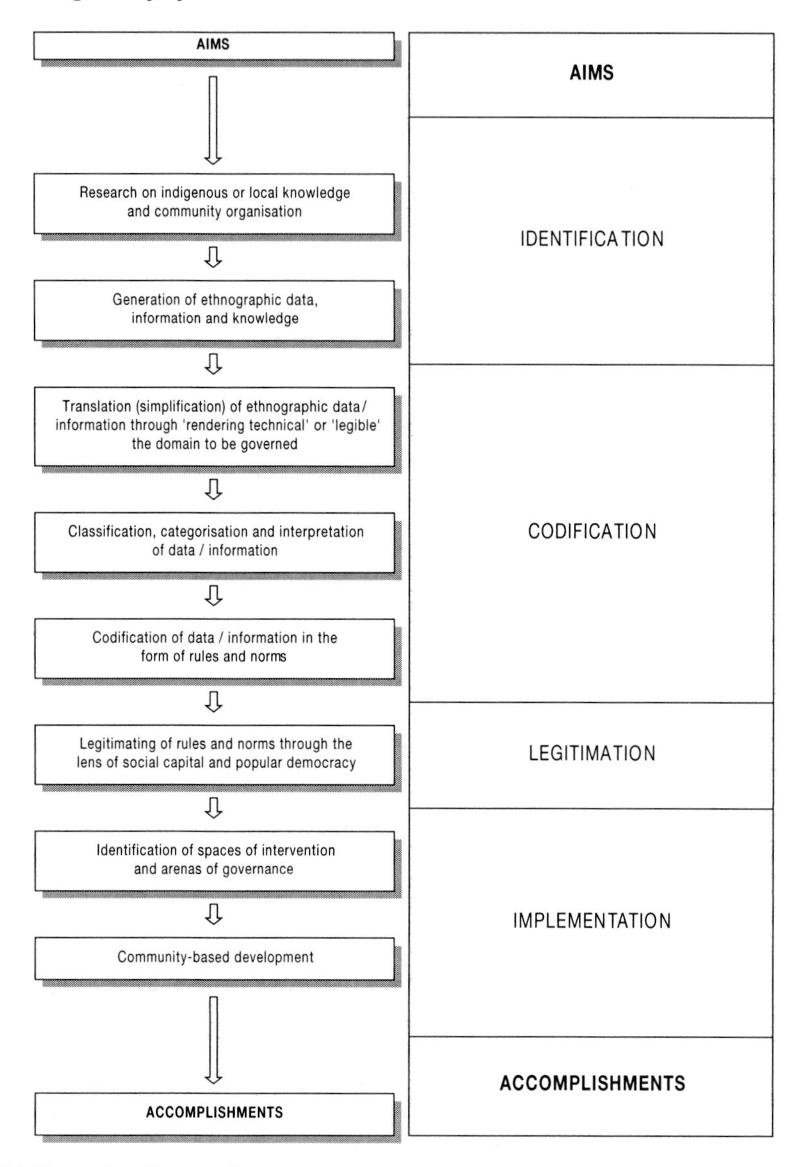

Figure 7.2 Governing the everyday

management and control. The state may, to use Scott's (1998) word, try to make the local 'legible' but it does this mainly to itself. There is much at the local level that the state is either unable to decipher or does not feel the need to decipher. The black box, so to speak, is capacious even when states are strong and systems of surveillance and control apparently well developed. Moreover, the way that the state deciphers the local may not be intelligible to local people. Third, and most importantly, there is the disjuncture between 'aims' and 'accomplishments', diagrammatically depicted at the top and bottom of Figure 7.2. This is where the truly everyday comes into play, creating a separation, often a chasm, between the aims of the most carefully constructed, orchestrated and well-funded project and what the aims actually achieve. This is not to say that such attempts at government and control

Table 7.5 Local, state and hybrid political geographies

Local	*Scholars highlighting the hybrid and in-between*	*State*
'Lower' level discourses	Bebbington et al. (2004) on Indonesia; Li (1999) on Indonesia	'Upper' level discourses
Society	Kerkvliet (2005) on Vietnam	State
The masses	Williams (1997) on West Bengal, India	The elite
Indigenous knowledge	Corbridge and Jewitt (1997) on Bihar, India	Expert knowledge
Informal	Hasan (2002) on Karachi, Pakistan	Formal
Resistance	See Chapter 8 on China (page 173)	Power

will always fail, or that local people are somehow insulated from the rule of the state, but simply to highlight the contingent nature of governance when viewed from the viewpoint of the everyday. Part of the reason why it is contingent is not because the state does not filter down to the local and the everyday, but because even when it does there are means, ways and strategies to resist what is unpalatable or unwanted. These 'everyday forms of resistance' are explored in Chapter 8.

A core point is that simply creating new spaces for political activity – whether through policies of decentralisation and devolution, through engendering participation and empowerment, through reworking citizenship, or through creating a network of local level groups and organisations – is often not sufficient fundamentally to alter the architecture of power in local areas. This is because the everyday drives a wedge between aims and outcomes. As a result, disjuncture is common, separation to be expected, and hybridity the norm (Table 7.5).

Further reading

The classic work on how states make people legible is Scott's (1998) *Seeing like a state*. Li's (1999, 2005) articles, however, provide a valuable corrective to the assumption that states are all-powerful, drawing on her ethnographic work in Indonesia. In recent years the participatory paradigm has come under attack from numerous quarters, reflected in the edited work of Cooke and Kothari (2001) and in articles such as those by Cornwall (2003) and Mohan and Stokke (2000). One particular theme has been the study of participatory exclusions and this can be very effectively traced through the India's Joint Forest Management programme (Corbridge and Jewitt 1997; Agarwal 2001). Two contrasting takes on the relationship between state and society can be found in Fan's (2001, 2002) work on the *hukou* system in China and Kerkvliet's (1995b, 2005) on everyday politics in Vietnam. The former shows how the state can closely control and regiment society; the latter how society can force through change to state policies from the bottom-up.

Making people and spaces legible

Li, Tania (1999) 'Compromising power: development, culture and rule in Indonesia', *Cultural Anthropology* 14(3): 295–322.

Li, Tania (2005) 'Beyond failed "the state" and failed schemes', *American Anthropologist* 107(3): 383–394.

Scott, James (1998) *Seeing like a state: how certain schemes to improve the human condition have failed*, New Haven, CT: Yale University Press.

Participation

Cooke, Bill and Kothari, Uma (eds) (2001) *Participation: the new tyranny?*, London: Zed Books.

Cornwall, Andrea (2003) 'Whose voices? Whose choices? Reflections on gender and participatory development', *World Development* 31(8): 1325–1342.

Mohan, Giles and Stokke, Kristian (2000) 'Participatory development and empowerment: the dangers of localism', *Third World Quarterly* 21(2): 247–268.

Case studies in participation

Agarwal, Bina (2001) 'Participatory exclusions, community forestry, and gender: an analysis for South Asia and a conceptual framework', *World Development* 29(10): 1623–1648.

Corbridge, Stuart and Jewitt, Sarah (1997) 'From forest struggles to forest citizens? Joint Forest Management in the unquiet woods of India's Jharkhand', *Environment and Planning A* 29(12): 2145–2164.

State–society, society–state

Fan, C. Cindy (2001) 'Migration and labor-market returns in urban China: results from a recent survey in Guangzhou', *Environment and Planning A* 33: 479–508.

Fan, C. Cindy (2002) 'The elite, the natives, and the outsiders: migration and labor market segmentation in urban China', *Annals of the Association of American Geographers* 92(1): 103–124.

Kerkvliet, Benedict J. (1995b) 'Village–state relations in Vietnam: the effects of everyday politics on decollectivization', *Journal of Asian Studies* 54(2): 396–418.

Kerkvliet, Benedict J. (2005) *The power of everyday politics: how Vietnamese peasants transformed national policy*, Singapore: Institute of Southeast Asian Studies.

8 Alternatives

The everyday and resistance

Introducing and defining resistance

Chapter 7 introduced the theme of resistance. This chapter will explore in greater detail how 'ordinary' people in the Global South resist and manipulate the forces that are arrayed against and around them. As James Scott (1985) writes in the preface to his book *Weapons of the weak*, rather than focusing on grand-scale, high-profile protests, rebellions and revolutions 'it seemed to me more important to understand what we might call everyday forms of peasant resistance – the prosaic but constant struggle between the peasantry and those who seek to extract labor, food, taxes, rents, and interest from them' (Scott 1985: xvi). Scott was writing about resistance in a rural context, drawing on some two years living in a village in the Malaysian state of Kedah between 1978 and 1980. Everyday resistance, however, need not be restricted to the peasantry but extends to all walks of life and arenas of work.

There are other characteristics of everyday resistance beyond its sheer prosaic nature. These include, importantly, its low visibility. Everyday resistance is often undeclared rebellion. It involves such actions and activities as foot-dragging and gossip. These are the so-called 'hidden transcripts' of resistance which have lain hidden from view and overwhelmed in people's minds and official texts by more overt forms of resistance.

The concern for everyday resistance or, as it is sometimes also known, subaltern (subordinate or inferior of rank) resistance, arose because of a wish to articulate and promote the 'voice from below' and to rewrite history from the perspective of the grassroots. This did not emerge only out of a concern that the emphasis on elite interpretations had produced a one-sided and distorted view of history. There was also a *political* desire to empower ordinary people through giving them a voice, taking their views seriously, and awarding them an agency denied in standard historical interpretations; to, in short, promote a 'politics of the people' (Guha 1982a: 4). The assumed authenticity of everyday/subaltern resistance also gives the perspective a legitimacy founded on its explicit links with the local and with tradition.

The attempt to write history from below has been most assiduously pursued in South Asia through the Subaltern Studies project. Drawing on the work on Antonio Gramsci, a group of young scholars banded together to pursue and develop the view that elite historical discourses in India had failed 'to identify, far less interpret, many of the most significant aspects of our [Indian] past' (Guha 1998: xv). This led to the production of ten volumes of essays between 1982 and 1999 and thirteen monographs. The first six volumes were edited by Ranajit Guha and the last four by an assortment of other scholars attached to the project.[1] Guha (1998: xiv), in a later essay, admits that the Subaltern Studies project was very much a child of its time and also notes that the scholars who led a project which inquired into the lives and significance of marginal people were, at the time, marginalised academics.

Defining resistance

Scholars have tended to conceptualise and define resistance in different ways (Table 8.1). In the context of this discussion, we can highlight four key fault lines. First, scholars differ in how they view the relationship between power and resistance; second, in terms of whether we can regard resistance as 'informal', while power/domination is formalised; third, with respect to the historical underpinning of resistance; and finally, with regard to the relationship between resistance and globalisation. These differences in emphasis and view should not be regarded as competing – so that one is 'right' or 'correct', and the others are not. Rather they represent the different ways in which scholars, often from different disciplines, have chosen to deploy the term resistance.

Resistance is increasingly used in the context of 'anti-globalisation' (Mittelman 2001: 214, and see below). Resistance becomes, almost by default, a reaction and a response to the manner in which people across the world are being drawn into global relations. The danger here is that such an approach tends to reify and celebrate the local as – almost by definition – progressive, authentic and in conflict with globalisation, and the global as reactionary and, by association, destructive or corrosive of local cultures, structures and livelihoods. As this book has attempted to show, everyday lives in the Global South are far more mixed and diverse than such a crude polarisation proposes. The reality is that much activity at the level of the everyday is not resistance *to* globalisation but resistance *for* globalisation. It is people's exclusion from the perceived benefits of globalisation as well as their struggle against the inequities of globalisation that inform resistance. We see therefore in ordinary people's resistance practices an often uncomfortable combination of actions and activities which set out – seemingly at one and the same time – to challenge, support, undermine, reinforce, stabilise and corrode existing power structures, hierarchies and processes. Resistance and domination are inherently ambiguous and ambivalent, and perhaps it is not least for this reason that scholars find the distinction problematic. That said, it does provide a valuable starting point for the discussion that follows (Table 8.2).

Table 8.1 (Everyday) resistance: defining terms

Theorist	*Definition/usage*
Karl Polanyi (1957)	Resistance to industrial capitalism arising out of its marginalising and polarising effects. The focus of resistance in Polanyi's work is on the formal and the official.
Antonio Gramsci (1971)	Resistance to the hegemonic roles of national (and international) institutions from the church to the family that give meaning to everyday life. Counter-hegemony movements may be overt and violent or low intensity and gradual.
James Scott (1985)	Resistance as 'infra-politics' – politics without attitude. '. . . [T]he persistent efforts of relatively autonomous petty commodity producers to defend their fundamental material and physical interests and to reproduce themselves' (Scott 1985: 302).
Michel Foucault (1990)	Resistance and power are viewed as co-constituted; resistance is a form of power and power begets resistance. Power tends to be coherent, aggregated and hierarchical; resistance is disaggregated, unstructured, decentralised and local.

Source: information extracted from Mittelman 2001

Table 8.2 Studies in everyday resistance and subaltern strategies

Region/country	*Source/context*
South America	
Latin America	Brass (2002a, 2002b) – critique of subaltern resistance scholars working on Latin American
Mexico	Barkin (2002) – on the persistence and perseverance of Mexico's 'culture of maize', and resistance to the challenges of globalisation, through engagement, diversification and delocalisation
Colombia	Jackson (1995) – on the politics of 'Indianness' and the invention of Indian culture
Asia	
China	Cai (2004) – on state-mediated resistance in China
India	Guha (1998) – essays drawn from the subaltern studies project which sought to write an Indian history 'from below'
Indonesia	Li (2000) – creating a politics of difference to confront hydropower plans in the Lindu Valley, Central Sulawesi
Malaysia	Scott (1985) – on everyday forms of resistance in the village of Sedaka, Kedah; Hart (1991) – on gender and resistance in the countryside
Malaysia	Ong (1987) – everyday resistance of female factory workers in Malaysia
Pakistan	Butz (2002) – resistance among porters (and trekkers) in Shimshal, northern Pakistan
Sri Lanka (Ceylon)	Duncan (2002) – strategies of resistance among Tamil plantation workers in colonial Ceylon
Thailand	Isager and Ivarsson (2002) – tree ordination and the power of religion to resist commercial logging and the territorialisation tendencies of the state in Thailand
Vietnam	Kerkvliet (1995a, 1995b, 2005) – resistance strategies employed by cooperative members against the cadres and the one party state in communist North Vietnam

Resistance and control in colonial and postcolonial contexts: four case studies

Resistance in nineteenth century Ceylonese coffee plantations

The plantation economies of the colonial world – in British India, Ceylon and Malaya, in the Dutch East Indies, in French Indochina, and in the American colony of the Philippines – were attempts to instil order and inculcate modernity in the tropical world. For Duncan (2002: 317), they were 'laboratories of modernity':

> Plantations can be conceived of as modern technologies for the reconfiguration of space, tools, scientific instruments and other material resources, bringing together culturally heterogeneous populations, stripping them of their former social attachments and reconstituting them as workers through the use of space-time strategies of monitoring the control.
>
> (Duncan 2002: 317)

There is little doubt that plantations did go further in reworking social and physical space in the interests of economy than had, hitherto, any other forms of agricultural production. Workers were transported thousands of kilometres to new territories and cultures, they were housed in 'lines' and their lives regimented and bureaucratised around the production needs of the plantation crop. Yet, even in such a context of strict surveillance, monitoring and control where workers were physically and emotionally dislocated from the social structures, networks and capital of home there was 'resistance'. Duncan (2002: 326–327) explains how Tamil workers in Ceylonese coffee plantations used multiple strategies to challenge the authority of the planter, exploit 'cracks' in the systems of control, and minimise their work: cry-off sick, slope off after muster and return to their huts, hide in the fields, lighten their loads, bamboozle the supervisor into thinking work had been completed, sabotage the pulping machine to catch a few moments rest, and steal coffee beans to sell in the market.

> Workers discovered ways to escape the planters' and overseers' panoptical procedures; to discover places that could not be seen; to learn how to take advantage of moments when supervision was lax; to manoeuvre . . . within an enemy field of vision.
>
> (Duncan 2002: 327)

These everyday forms of resistance were ongoing and continuous – part of the day-to-day landscape of plantation production. It was a game played between protagonists where the rules or limits were broadly acknowledged, even if these rules were not articulated. At times, of course, everyday resistance escalated into more overt forms of resistance that took the form of open insubordination, recourse to the courts for the payment of wages, or even desertion. But this escalation was unusual and, in a sense, marked the failure of the hidden transcripts of everyday resistance (and control) to permit systems such as Ceylonese plantations to muddle along.

Gender and resistance in twentieth century factories

The plantation can be seen as the forerunner of the industrial factory in the way that rural people ('peasants') were moulded and absorbed into modern systems of production relations. At several points in this book the struggles of women have come to fore, particularly in the context of women and factory work (see page 108), women and mobility (see page 129) and women and the household (see page 44). Women's resistance in the context of the factory environment is not just a struggle against the forces of capital; nested in such strategies are reflected parallel struggles in wider society. Thus, for Hart (1991: 95), women's resistance in the factory is not just an issue which intersects with production relations on the shop floor but one which resonates with gender relations in the household and the community.

It has become usual to view export-oriented factories in the Global South as sites of surveillance and control where young women are drawn into regimented, industrial environments which sometimes – particularly in Asia – reproduce the patriarchal context of the household. Earlier chapters have already challenged this simplified and simplistic vision of 'the factory' in pointing out that factory work can be empowering, giving unmarried young women a degree of self-determination, a modicum of income, an improved status within the household, and a heightened sense of self-worth (see page 111). These arguments, though, tend to focus attention on the worker's place in the household not on the worker's place in the factory. The question therefore arises whether the potentially empowering effects of factory work vis-à-vis household relations are accompanied by domination and exploitation in the factory itself.

In the export-oriented electronics factories of Malaysia we see, reflected in women's subaltern resistance strategies, an intersecting and sometimes contradictory set of gender norms and ideologies:

- pre-colonial representations of women as active and financially astute;
- modern representations of women as housewives and mothers;
- contemporary Muslim representations of women, arising from the Islamic revivalist (*dakwah*) movement, as confined to the private and domestic spheres;
- and modern representations of women as willing members of the industrial workforce driving Malaysia's modernisation project.

Aihwa Ong sees protest in electronics factories in Malaysia arising not due to the emergence of a nascent class consciousness but as an outcome of the 'violation of one's fundamental humanity' (Ong 1987: 202). The young women were embedded in capitalist production relations but, in terms of their self-definition, were non-capitalist *orang kampung* (village folk). How, then, do simple country folk resist the machinations of capital? They do so – metaphorically, and sometimes literally – by throwing a non-capitalist spanner in the works. But, as young Malay women who are expected to be *takut dan malu* (fearful and shy), they do so in ways that are covert rather than overt, disruptive rather than destructive, and traditional rather than modern. They leave the factory floor citing 'women's problems', slow down their pace of work, spend longer periods in the prayer room, or become shoddy in their working practices (Ong 1987: 203).

More dramatically, there have been numerous reports of mass hysteria caused by spirit possession among women factory workers in Malaysia's export oriented factories. Traditional spirit healers or *bomoh* are called in to purify the factories and exorcise the spirits or ghosts. For modern managers such actions are irrational hangovers from a pre-modern past. For Ong, however, spirit possession is a form of resistance against 'labor discipline and male control in the modern industrial situation' (Ong 1987: 207). Spirit possession challenges the rules and disciplines of modern factory work, offering a hidden strategy of revolt in a non-unionised world:

> the inscription of microprotests on damaged microchips constituted an anonymous resistance against the relentless demands of the industrial system. These nomadic tactics, operating in diverse fields of power, speak not of class revolt but only of the local situation.
>
> (Ong 1987: 213)

Religion as resistance: tree ordination in Thailand

The territorialisation efforts of the state have already been discussed in Chapter 7 (see page 146). However, this focus on the ways in which the state makes people and places legible has also led to a growing interest in processes of counter-territorialisation. Counter-territorialisation involves efforts by local people to resist, challenge and subvert the territorialisation tendencies of the state through a variety of grassroots efforts from counter-mapping to tree ordination. Sometimes this is individualised; sometimes the efforts are articulated through people's organisations; and sometimes they are enacted with the assistance of NGOs.[2] One such effort has been the tree ordination movement in Thailand.

From the early 1990s, Buddhist monks with the support and assistance of environmental activists and NGOs, particularly in northern Thailand, began to 'ordain' trees by encircling their trunks in saffron-coloured cloth – a process known as *buat paa* (Illustration 8.1). While there has been a long tradition in Thailand of honouring certain trees, particularly the Bodhi tree (*Ficus religiosa*) under which the historic Buddha attained enlightenment, ordaining non-sacred trees was something entirely new. Isager and Ivarsson (2002: 404) believe that the first act of tree ordination in Thailand occurred in 1988 when the abbot of Wat Bodharma in Phayao Province, Phrakhru Manas Natheepitak, used it as a strategy to confront logging in the watershed forest close to his monastery. Since then it has become a common strategy by which relatively powerless local people can tap into the power of religion to thwart the state and state-supported or state-linked business interests (such as logging companies). From being regarded as slightly weird in the early 1990s it had become, by the end of the decade, a mainstream and broadly accepted practice.[3]

Isager and Ivarsson (2002) recount a defining experience in Mae Chaem district in Chiang Mai province in 1993 where the villagers of Ban Chan ordained 1,000 pine trees to prevent their logging by the Forest Industry Organisation. Since then, many other villages, faced with the prospects that their community forests might be logged, or concerned at the denudation of watersheds have used the strategy of *buat paa* to resist the machinations of the state and capital. Tree ordination is, in itself, a remarkable example of an everyday weapon of the weak. It seems, simultaneously, to pay homage to several powerful strands in the localist agenda. It is rooted in local belief structures. It is a traditional strategy which has been effectively used to confront modernising influences. It pits the non-commercial against the commercial. It is participatory. It is generated and driven by local people according to local needs and concerns. And it draws on religious symbolism and imagery to confront a secular, managerialist and technocratic business and political world.

Illustration 8.1 The power of religion – 'ordained' trees in northern Thailand

But there is both more and less to tree ordination than this list suggests. The forested uplands of northern Thailand are contested landscapes. This contestation is not just between local people and the state, or even between local people and the state plus business interests. The uplands are inhabited by a mix of hill 'tribes' and lowland Thai so we see in the hills of the North a multifaceted struggle between different interest groups that cannot be easily categorised as 'local' and 'non-local'. Isager and Ivarsson (2002) note that it is ironic that *buat paa* is usually classified as a counter-territorialisation strategy because it is, in fact, an alternative territorialisation strategy. Local people, farmers, Buddhist monks, NGOs workers and environmental activists are using adapted religious practices to territorialise land to challenge mainstream territorialisation tendencies. It could also be argued that far from being a strategy that runs counter to the tide of modernisation and development, many such strategies conform to the broad direction of change. Local people ordain trees so that they can secure resources for their own productive use – and to exclude others, such as hill peoples, who might claim rights to those resources. To return to a point made in the opening section to this chapter, 'resistance' cannot simply be read-off as a challenge to established practices and power structures. Tree ordination practices in northern Thailand, for example, often reflect local level conflicts and struggles rather than local–elite or local–state conflicts.

Everyday resistance in one party states: Vietnam and China

People living in democracies can make recourse to what is sometimes termed non-institutionalised action through such bodies as new social movements.[4] But what avenues are open to people in authoritarian contexts where the space for such bodies to operate is circumscribed and curtailed? It seems that in such circumstances, there are two main ways in which resistance can occur. First of all, resistance becomes concentrated at the everyday level. This is explored below with reference to Vietnam. Second, elements of resistance may be transmitted – and intentionally so – through official channels and the structures of the state. Such state-endorsed resistance is evident in the way in which the central authorities in China manage and manipulate local level dissatisfactions.

During the 1960s and 1970s the Communist Party in northern Vietnam, as it struggled to meet the needs of a growing population, 'implored' peasants to devote more time to collective work, setting a target of 250 days per year for each adult worker. With considerable cajoling, the days of collective work did, indeed, increase reaching 236 days by 1975. However, as Kerkvliet explains, these were notional days of collective work. He quotes one respondent saying that labour was 'a few hours each day and a little work each hour' (Kerkvliet 2005: 109). Corners were cut, work was shoddy, records were doctored, and officials were deceived and hoodwinked. Instead, peasants in Vietnam devoted enormous efforts in time and energy to the cultivation of their small private plots and in other forms of private enterprise. Cooperative members played the system, 'hiding' land from the authorities, fixing contracts (what became known as *khóan chui* or 'sneaky contracts'), and siphoning off grain from collective harvests to private stores. These everyday forms of resistance meant that while notional hours devoted to collective work were increasing, net income from such work was actually declining so that by 1971 just 30 per cent of the average cooperative member's net income came from collective work (Kerkvliet 2005: 114; see also Kerkvliet 1995b: 405) (Table 8.3). Ho Chi Minh may have encouraged 'everyone to work as hard as two' but peasants, disgruntled with their cooperative and party leaderships and disenchanted with the lack of incentives in the collective system, ignored the plea and instead turned Uncle Ho's dictum into a mildly subversive ditty:

Everyone work as hard as two
so that the chairperson can buy a radio and bicycle.
Everyone work as hard as three
so that the cadre can build a house and courtyard.
(quoted in Kerkvliet 2005: 111)

Ultimately these everyday actions fundamentally undermined the logic of the collective system causing it, between 1974 and 1981, to 'collapse from within'. 'Disinterest and disgust were so serious in some areas that tens of thousands of hectares went unplanted' (Kerkvliet 1995a: 69).

Table 8.3 Collective work, individual returns in North Vietnam, 1959–1975

	Days allocated to collective work	*% contribution from collective work to average cooperative member's income*
1959	90	38 (early 1960s)
1970	215	30 (1971)
1975	236	

Source: data extracted from Kerkvliet 2005: 109 and 114

In China, like Vietnam, such low visibility and everyday forms of resistance are prevalent. They are important given the fact that strikes are illegal and demonstrations require prior approval from the authorities. Nonetheless, the most common form of overt citizen action in China is for individuals or groups to bypass the local level tier of government and take their grievances direct to higher levels. One survey carried out among 1,461 respondents in the cities of Beijing, Harbin, Wuhan and Guangzhou revealed that approaching such higher level authorities accounted for 67 per cent of actions (Cai 2004: 429). This appeals system is important in China because, without a free press and with a weak legal system, power and control are concentrated within the local level administrative hierarchy. In 1999 collective appeals in the five provinces of Henan, Shandong, Guangxi, Jiangsu and Jilin alone totalled 66,417. The centre uses the appeals system to monitor citizens' views and gauge the effectiveness of local level administration – and to take action against abusive local officials when necessary. A large number of appeals issuing from one area is taken to indicate that local officials are not fulfilling their responsibilities (Cai 2004: 438).[5]

The case of China raises questions about the resistance/domination binary noted in Chapter 7. Effectively the state is managing citizen participation through a form of mediated resistance in which the centre creates a space of resistance where disgruntled people can register their dissatisfaction with local level officials and units. From the centre's point of view, of course, it is attractive because it keeps resistance within the ambit and the control of the state – rather than permitting it to leak into the private arena and become potentially destabilising.

Resistance as anti-globalisation?

Although, as noted in the introduction, resistance should not be seen as synonymous with anti-globalisation there is no doubt that this is an important thread in the literature. Chapter 5 explored the impact of the development and spread of global agro-food systems on livelihoods in Mexico (see page 100). This has also involved, it has been argued, a resistance response by which rural communities 'construct their own social and productive alternatives to respond to the challenges of globalization' (Barkin 2002: 82). To sustain the rural maize economy in Mexico, people are migrating to other areas, remitting money, and subsidising and sustaining the rural economy and rural production through engagement with the non-local and the non-rural. Ironically, then, resistance in Barkin's interpretation of Mexico's 'culture of maize' comes not from the relocalisation of production (i.e. emphasising traditional maize production) but from delocalising employment, work and livelihoods *so that* traditional cultures can be strengthened.

This paradoxical combination of the local and non-local in subaltern strategies of localisation is also explored in Escobar's (2001, 2004) work on the Pacific rainforest region of Colombia. Here, the progressive opening up of the Colombian economy to the world economy from 1990 has brought together forces of delocalisation and localisation (although he also emphasises that these are not bounded and discrete). Among the former are global capital in the form of African palm plantations and industrial shrimp production, and the techno-science of global conservation movement; among the latter are place-based social movements and local conservation initiatives.

Place-based strategies enact a politics from below, but this is implicated in networks and relationships that stretch beyond the local. Thus activists of the Process of Black Communities employ a political ecology framework that links their place-based movement not only with local communities but also with NGOs, the media, academics, and the state. Traditional production practices, indigenous approaches to biodiversity, cultural conceptualisations of territory, and locally rooted ethnic identities are 'interwoven by movement activists into a discourse for the defense of place and a political ecology framework that enables them to articulate a political strategy' (Escobar 2001: 163). Escobar sees this operating as a threefold strategy of localisation. First, a place-based emphasis on local cultural forms and production practices. Second, a trans-local strategy which engages with other and wider movements, be they environmental, cultural or anti-globalisation. And third, a political strategy directed at making explicit the links between identity, territory and culture.

Arturo Escobar (1995) has taken this localist resistance agenda further and wider in his influential book *Encountering development: the making and unmaking of the Third World.* He argues that since the mid to late 1980s, grassroots movements across the Global South have begun to offer an alternative to mainstream development discourses and approaches. Escobar writes:

> In spite of significant differences, the members of this group share certain preoccupations and interests: an interest in local culture and knowledge; a critical stance with respect to established scientific discourses; and the defense and promotion of localized, pluralistic grassroots movements.
>
> (Escobar 1995: 215)

For Escobar, this movement has emerged *in opposition* to capitalist development. It seeks to present not an alternative development, but an alternative to development. It is a political movement and, invariably, it is a movement against globalisation and all globalisation is seen to stand for.

In manufacturing effective resistance strategies, local people often draw upon non-local resources and expertise. This is most usually in the form of the support and advice offered by NGOs. In the case of the village of Shimshal in northern Pakistan, resistance to efforts to impose a national park on the traditional territory of the village led to a discursive battle over environmental stewardship.[6] In 1975, the 2,300 square kilometre Khunjerab National Park was established and designated as a World Conservation Union (IUCN) Category II protected area. The Shimshalis were portrayed by external agencies as poor stewards of the environment, inherently destructive, thereby creating a space for external intervention and externally imposed environmental management. Realising the way in which they were being discursively constructed as a threat to the integrity of the environment, some younger and more educated villagers created their own, competing, Shimshal Nature Trust (SNT), which set formal guidelines for the sustainable management of 2,600 square kilometres of land. As with several of the other case studies presented in this chapter, central to the Shimshalis efforts is an explicit attempt to link land (territory) with their collective culture, history and identity. To communicate their efforts to the wider world, they utilised an academic to put their supporting document and justification – a 'Fifteen Year Vision and Management Plan' – on the web (http://www.brocku.ca/geography/people/dbutz/shimshal.html). The rationale for the SNT is stated to be:

> Until recently we have not felt the necessity for a formalised nature stewardship programme. Four hundred years of sustainable interaction with our landscape offered ample proof of the sustainability of community members' environmental practices. In the past decade, however, progressively greater access to, and interaction with, the outside world has threatened to both alter our community's traditional relationship with nature, and to remove control of that relationship from the community.
>
> The effort to develop a Shimshal Nature Trust is also a response to our experience with Khunjerab National Park (KNP) . . . [which covers] most of Shimshal's pastoral territory, as well as the communal pastures of eight other villages. Other affected communities have been willing to accept (but have not yet received) compensation for their loss of access to traditional pastures. We alone are unwilling to relinquish access to our pastures under any circumstances, a position we justify by . . . outlining our community's historical and current symbolic attachment to parts of the territory under threat. We were not consulted in the delineation of the park boundaries, in the definition of park regulations and land-use restrictions, or in the details of park management.
>
> While we appreciate recent efforts by external agencies to develop community-based nature conservation projects [but] suggest that it is not enough that external initiatives be managed locally; rather, a culturally and contextually-sensitive nature stewardship programme should be developed and initiated, as well as managed, from within the community. In keeping with this emphasis on local context, we have decided to build our Shimshal Nature Trust around a broad definition of environment, which includes socio-cultural and ecological components in relationship with each other.
>
> (http://www.brocku.ca/geography/people/dbutz/shimshal.html)

This short extract from the KNP documentation makes mention of a number of issues central to many subaltern resistance strategies:

- the contrast between local and external actors
- the wresting of control from local communities by outside agencies
- the value of local/indigenous knowledges as against external/elite knowledges
- the role and efficacy of community-based management
- the historical, cultural and symbolic links between people and territory.

For Butz (2002: 27), the experience of Shimshal is an example of the 'careless globalisation of environmentalism and sustainability discourses'. But what is significant is that rather than combating this in traditional ways, the Shimshalis' methods of resistance have utilised global media technologies so that they can 'disrupt the globalising flow of power represented by organisations like the IUCN' (Butz 2002: 27). The Shimshal and other examples in this section demonstrate that resistance is rarely about globalisation alone. Nor is the local/global binary often particularly revealing in explaining why resistance arises, and what forms resistance can take.

The anti-globalisation movement is often presented as 'new' – something that emerged and solidified over the 1980s and 1990s. Just as there is a case for tracing the origins of globalisation back to the fifteenth and sixteenth centuries,[7] so anti-globalisation can be seen to have deeper historical roots, or antecedents, than usually accepted. Broad and Heckscher (2003; see also Sadler 2004) identify three periods of resistance linked with earlier waves of globalisation – namely, the anti-slavery and international workers movements associated with European colonialism, the import-substitution industrialisation strategies of the early post-Second World War period (1950s and 1960s), and the attempts to forge a New International Economic Order during the 1970s. They suggest, furthermore, that the parallels between today's anti-globalisation movement and their antecedents are 'striking' and that the current movement has 'much to learn from [this] resistance to earlier forms of coercive economic integration' (Broad and Hecksher 2003: 724–725). Two significant differences between these earlier anti-globalisation movements and today's initiatives are, however, the degree to which both the targets and the sources of resistance are different (see Sadler 2004: 854). Rather than targeting nation states, the anti-globalisation movement today tends to mobilise against corporations and multilateral organisations; and the agents of this resistance are NGOs and people's organisations which are either based and mobilised from the grassroots or, at least, take their inspiration and gain their legitimacy from the grassroots.

A key issue over which there is ongoing debate – and disagreement – is whether new social movements, and the place-based political movements of which they are often a part (see the next section), can be interpreted as representing 'anti-globalisation'. Escobar (2004: 209–210) argues that because there are no modern solutions to many of today's modern problems this requires scholars to 'imagine beyond modernity' for answers. He uses the example of Colombia to make it 'patently clear the exhaustion of modern models' (Escobar 2004: 216) and, drawing on the struggle of black communities of the Pacific, highlights and lauds the possibility for locally rooted social movements to think and act beyond modernity and beyond the Third World (as he puts it). He calls this counter-hegemonic globalisation in the sense that locally rooted movements draw on trans-national networks for their vitality. Escobar's views can usefully be contrasted with the work of Antony Bebbington (2004a) on federations of indigenous Quichua communities in the central Andes of Ecuador. For Bebbington, such movements can be successful only if they make a material difference to livelihoods and contribute to delivering the basics of employment, income, and increased productivity. He is sceptical of the view that new social movements are a form of resistance to globalisation and argues that it is by delivering sustainable livelihoods – which often requires a deeper engagement with the market and market relations – that Indian identities can be maintained and fostered. He then shows how increased commercialisation has been accompanied by the maintenance of cultural identity in terms of dress, language, kinship and so forth. Bebbington, in effect, challenges most of the assumptions embedded in Escobar's work. For him, this 'bottom-up modernisation' among the Quichua has led to a rejection of traditional technologies because they are associated with the Quichua's subjugation through the hacienda system.

Modern technologies are, therefore, seen as politically progressive, socially empowering, and community preserving:

> Quichuas in Chimborazo do not have time to wait for the dawn of new utopias. They demand liberation from where they are now. The challenge then is to build short-term, pragmatic, and realistic responses that work from contemporary contexts, and do so in a way that is coherent with and builds towards longer-term utopias that are already immanent within the strategies and hopes of popular sectors.
>
> (Bebbington 2004a: 415)

Earlier in this chapter it was noted that while new social movement may represent the politics of a class, they do not pursue politics on behalf of a class (see note 4). We can develop this a little further and state that new social movements in the Global South often serve to ferment and pursue the politics *of a place*. It is to this issue which the chapter now turns.

The politics of place and the politics of difference

Many local level political movements and resistance strategies, such as those lauded by Escobar, are connected with engendering and promoting a politics of place. The politics of place is also often closely linked to the politics of difference and, therefore, to cultural politics. Indeed, one effect of the concern with everyday resistance has been to re-energise 'place'. Social movements – defined as the collective efforts of people united by a common purpose – are, more often than not, highly place-sensitive. Likewise the relocalisation turn in geography, anthropology, environmental studies and related disciplines has encouraged scholars and activists to reconsider the role of place (Escobar 2001, 2004; Elmhirst 2001; Oslender 2004). Politics is not, however, just a reaction and a response to place; it also creates places by endowing them with cultural value and historical significance. A case study of the politics of 'Indianness' in Colombia is provided in Box 8.1.

BOX 8.1 The politics of Indianness in Colombia

It is sometimes the case that everyday resistance is based on the construction, creation or re-creation of an indigenous identity. Jean Jackson (1995) explores just such a process in connection with the emergence of a sense of 'Indianness' among the Tukanoans who live in the Vaupés region of south-east Colombia. (For another discussion of the role of culture in development among indigenous peoples of the Andes see Radcliffe and Laurie 2006.)

Jackson argues that

> if Tukanoans are to have any power at all [in the context of a highly bureaucratised and centralised state] they *must* have a traditional culture. Winning the battle for self-determination increasingly involves acting and speaking with an authority that arises from an 'Indian way'.
>
> (Jackson 1995: 5, emphasis in original)

The Colombian state awards indigenous people special rights only if they can demonstrate their cultural distinctiveness. In addition, NGOs often target their attentions and their resources at those people who can demonstrate their Indian identity. Thus the Tukanoans are not so much resisting mainstream society but negotiating among themselves, sometimes with the support and input of other activist groups, about what forms Tukanoan culture should embody so that they can claim a right to difference and, in so doing, also claim the special rights that come with difference.

In Colombia, as in many other countries, it is the wider political arena within which indigenous peoples are located which has created and stimulated the emergence of a politics of Indianness. For this reason, 'Indianness' among the Tukanoan has more in common with established worldviews about Indians and non-Indians in the forests of the Amazon, than it has with Tukanoan cultural identities. A 'simplified, romantic and idealized image of Tukanoan society and culture' has, in the process, been created (Jackson 1995: 15). Cultural identity has been closely associated with land and territory when it seems that traditionally there was only a loose association between land and Tukanoan identity.

Being Indian involves, in the Tukanoan case, a careful tiptoeing between positions. For example,

> CRIVA (the Regional Indigenous Council of the Vaupés) members must refer respectfully to the past without appearing reactionary; they must promote and seek progress without appearing to sell out Tukanoan uniqueness or buy into assimilationism; and they must champion Tukanoans' right to a place in the sun in a multiethnic society without appearing to endorse overly separatist policies.
>
> (Jackson 1995: 12)

CRIVA has been awarded the goal of 'defending the culture[s]' of the Vaupés but to do this, of course, there has to be a culture to defend. There is also a further paradox wrapped up in CRIVA's mission (and also that of many NGOs) which is that because most development efforts promote change they will also lead to cultural loss – which CRIVA is enjoined to 'defend'.

Source: Jackson 1995

The Lindu people of the Indonesian island of Sulawesi began to coalesce as a political force in the late 1980s when a hydropower scheme was proposed for Lake Lindu.[8] The project became the stimulus – and the rationale – for an 'awakening' of Lindu cultural identity. With the support and active involvement of NGOs, the Lindu began to produce documents detailing their unique culture and language, stressing their historical and emotional attachments to their lands and territory, outlining the negative impacts of the dam, and stressing their tribal (but non-primitive) identity (Li 2000: 163–168; see also Li 2001). The state did not acknowledge the 'special' nature of the Lindu, and continued to treat them as ordinary rural folk for whom the dam would bring development and whose livelihoods could simply, through judicious planning and investment, be reproduced elsewhere. The Lindu, on the other hand, were intent on creating an image of themselves as special and unique with an attachment to place that made their resettlement necessarily destructive. For the Indonesian government, the land issue is generalised; for the Lindu, it is spatially particularised. It is not that the loss of land that is the issue; it is the loss of *this* land.

In the case of the Lindu, the landscape of resistance pitted local, 'indigenous' people against the state in a vertical struggle. Struggles can also take a more horizontal form as different groups within localities compete with one another. This was explored in the context of migrant identities in Chapter 6 (see page 133). In Papua New Guinea's West New Britain province land shortages have created a resource context in which identities are forged and then harden, accentuating differences between settler and landowning groups (Koczberski and Curry 2004: 362). A still more intriguing study of resistance and the politics of place and scale is Butz's work on the village of Shimshal in the mountains of northern Pakistan, introduced in the previous section.

Traditionally the mainstay of Shimshal's economy was livestock rearing; nowadays, however, a significant proportion of income comes from taking on portering work for foreign trekkers and climbers. To be a porter in Shimshal is one of the few ways in which a man can earn a living. At the same time, being a porter is far from seen as a desirable occupation. The porter is very clearly subordinate to both trekkers and tour leaders, they are subject to oppressive labour relations, susceptible to fraud and exploitation, and face the physical risks that come from hard physical work in difficult and dangerous conditions. But, as Butz (2002; see also Butz 1995) points out, young men in Shimshal have little choice but to accept work as porters; alternative opportunities are simply unavailable so avoiding portering is not a viable option for most households. Like Tamil plantation workers in colonial Ceylon, porters in Shimsal employ various resistance strategies to reduce their work load and increase their return. They feign sickness, lighten their loads, set a slower pace, even sabotage the trek or desert all, as Butz (2002: 19) says, 'to wring less labour, and more money and control out of the portering encounter'. But while such everyday forms of resistance may, in the short term, produce the desired effects – more money for less work – they also, in the long term, create a reputation for the Shimshalis among foreign tourists and domestic tourist agencies of 'greed and recalcitrance'. This reputation has the effect of reducing the number of tourists coming to Shimshal, thereby eroding a livelihood activity upon which men in the village are increasingly dependent.

There is, therefore, an essential contradiction between the Shimshalis generally effective resistance strategies and their longer-term livelihood interests. Porters have tried to deal with this essential contradiction, Butz argues, by 'enlisting trekkers as active co-conspirators, by attempting to sell practices of resistance as a struggle for "authenticity" in which trekkers can participate, in opposition to guides, tour companies, and their own material interests' (Butz 2002: 22). Those younger Shimshal with some English will regale trekkers with accounts of Shimshal history and culture, drawing the tourist into a constructed arena of resistance of which they – the tourist – can become a part. For Butz (2002: 24), the trail is a 'third' space, a space which is global and local, inside and outside, indigenous and metropolitan. The trekker and the porter are brought together for a shared moment of resistance.

There is sometimes a tendency to think and write about ethnic 'groups' and indigenous 'people' as if common origins equates to shared interests. What is clear, and this point has already been made several times but in different contexts, is that the labels we apply to people – and they claim for themselves – may disguise deep-seated differences. Resurreccion's (2006) study of the Kalanguya, an ethnic group from the highlands of the Philippine Cordillera, is a case in point. The Kalanguya are traditional swidden agriculturalists where women grow sweet potatoes on cleared hill slopes and men hunt in the surrounding forests. This traditional livelihood practice has not been sustainable, however, for some years and today men, taking advantage of improving communications and rising levels of market integration, grow paddy rice and vegetables for sale while women work as labourers on men's fields or engage in trade. The 1990 Philippine Forestry Master Plan, in a similar vein to such efforts elsewhere (see page 152), accepted that local people should have a central role in community-based forest management initiatives and 'indigeneity has therefore become a powerful symbolic and state-recognized resource with which to stake claims, and claim authority, on resources' (Resurreccion 2006: 381–383). It is through claiming indigeneity that the Kalanguya can claim territory. So far, so familiar. Where Resurreccion's study diverges from the normal line of argument is in revealing how men and women have positioned themselves very differently in terms of the opportunities that have arisen from the introduction of community-based forest management. Men's role has been reconfigured as that of primary agriculturalists. Women, on the other hand, have seen their traditional farming role usurped. For men, the logic of claiming indigeneity in order to stake a claim for land and territory is therefore all too clear. Women, however, have tried to distance themselves from their traditional label of 'swidden cultivators' so that they can more easily pursue occupations in retail trade and wage employment. Men were happily using women as 'ethnomarkers' of sustainable development by, for example, 'calling attention to women's traditional erosion-control technologies in order to strengthen their political claims on ethnic identity and ancestral land' (Resurreccion 2006: 392).

> Kalanguya women and men demonstrated very different ways of managing and inserting themselves into the modernization process. Men invoked the past, identifying themselves as indigenous, in order to coalesce with the state in the present. Women rejected the past, silencing their associations with indigeneity, in order to fulfil their roles as food managers and link with the wider economy.
>
> (Resurreccion 2006: 395)

These examples illustrate that there are several 'cuts' that we can make through local political action (and inaction), which often intersect (Table 8.4). Resistance is usually characterised using one of these political cuts but, in many cases, there are multiple underpinning

Table 8.4 The multiple politics of the local

Politics	*Description*	*Example*
Politics of place	Place-based politics, and place-creating politics	The Lindu people of Sulawesi's cultural awakening and emergent place-based identity stimulated in response to state-led hydropower plans (Li 2000)
Identity politics	The politics of difference	Indigenous/settler conflicts over land in West New Britain province, Papua New Guinea (Koczberski and Curry 2004: 362)
Politics of scale	Global versus local; society versus state	Shimshal porters and trekkers shared resistance in northern Pakistan (Butz 2002)
Cultural politics	Politics based on cultural constructions of identity	The politics of Indianness in Colombia (Jackson 1995); gender and identity among the Kalanguya of the Philippines (Resurreccion 2006)

factors at work so that resistance which at one level simply appears to be a (possibly unequal) contest between poor peasants and the state, or workers and capital, may also be about local resource struggles, household gender relations, or inter-household livelihood tensions (see Elmhirst 2001).

Critiques of resistance and subaltern studies: a conclusion

As one would expect, subaltern studies and resistance strategies have not escaped the scholarly cauldron unscalded. Indeed there has been a sustained critique of subaltern studies. This, however, has been directed not so much at what scholars and activists have tried to do – to turn academic attention away from elite visions of history and society – but rather the assumptions that have underpinned this reorientation.

To begin with, some critics have questioned the 'authenticity' that is often applied to grassroots consciousness. This leads to a romanticisation of the past and of tradition which, in turn, leads to an uncritical lauding, celebration and promotion of peasant cultures and economies (Brass 2002b). Tradition is essentialised and set in opposition to modernity. Those who would like to see revolutionary change in the Global South also regard such approaches as reactionary and highly conservative in political terms because they celebrate peasant cultures and thereby, perhaps unintentionally, endorse the class relations which produce and reproduce the peasantry. Change requires a differentiated peasantry.

A second area of concern focuses on the way in which 'workers' and 'peasants' are awarded an agency which overlooks the confining contexts of class and gender relations within which agency must operate. They are far from being 'free-willed individual decision makers' (Hart 1991: 116) but are often boxed in and constrained by prevailing social structures, economic resources and political events. Just as workers and peasants are automatically allocated to a homogenised and problematic 'subaltern' category so – and this is the third area of critical concern – they are also assumed to 'resist', while elites dominate. As Ranajit Guha writes in the preface to the first volume of essays from the Subaltern Studies project, 'subordination cannot be understood except as one of the constitutive terms in a binary relationship of which the other is dominance' (Guha 1982b: vii). Most subalterns, however, are both dominated, and dominate; and similarly for elites. Rather than ascribing particular positions to two unwieldy and differentiated groups, we should really be identifying and examining the circumstances and contexts within which domination and resistance occur. The challenge, then, lies in accepting that subalterns and elites are factionalised social categories, that domination and resistance are often co-constituted, and moving on from these positions to assess how, why and when domination/resistance arise.

The fourth critique involves the way in which acquiescence or complicity are, often, more usual and prevalent than resistance. To view the Global South as a stage where elite classes, global capital and the forces of modernity are arrayed against ordinary people and their interests, and where the latter by their wit and wisdom cunningly resisting the domination of the former, underplays the degree to which ordinary people rework or work with the forces of modernisation. The domination/resistance binary, as noted in the last chapter, simplifies a set of relationships with multiple axes.

A final and fifth source of concern is the tendency for 'resistance', even when it is classified as 'everyday', to pay excessive attention to points of friction and tension – and thereby overlook the normal patterns of activity that lie beyond the field of resistance. Li uses the example of the Lauje of upland Sulawesi in Indonesia (Li 2000: 162–163) as a case in point. The Lauje do not feel pressured by the state. Their lands are not at threat of being taken by the state or by big business. They do not feel the need to adopt a collective position in order to challenge outside forces. They are not engaged in a struggle against the state, capital or the wider development project. They are, to be sure, cynical of the state and its capacities and intentions but Li finds it difficult to see any sort of coherent resistance movement emerging among the Lauje. For this reason she prefers to write of '"everyday" patterns of action and inaction' (Li 2000: 162), rather than everyday resistance.

Reflecting on the examples, case studies and debates which this chapter has raised, and it becomes clear that things are not always – or even usually – as they appear. Young women from rural backgrounds are not powerless. Counter-territorialisation strategies work within, rather than without, standard practices. Subaltern strategies are as normative as those mainstream strategies which they aim to challenge. Studies of everyday resistance often attempt to identify and trace the lives of individuals for the reason that subaltern studies, at its core, is about people. As Mallon (1994: 1507) admits in her consideration of subaltern studies in Latin American history, 'I . . . want to touch the pictures of the historical subjects I struggle to retrieve' and yet she also knows that there are no real individual subjects to be found. The archives, as she says, refuse to yield clear pictures. This is nearly as true for those scholars who explore the present, as those who mine the past.

Further reading

An interest in and concern for subaltern studies and everyday forms of resistance can be traced from the work of the Subaltern Studies project in India. Ranajit Guha (1998) has edited a Subaltern Studies reader which provides a valuable background while Scott's (1985) book *Weapons of the weak* is the fullest attempt to look in detail at how the poor and marginal (in this case, in Malaysia) make their voices heard. For further material on the case studies of resistance in this chapter see Duncan (2002) on workers' resistance on the coffee plantations of colonial Ceylon, Isager and Ivarsson (2002) on tree ordination in Thailand, and Ong (1987) on factory work and resistance in Malaysia. General reviews of resistance to globalisation – or anti-globalisation – are provided by Mittelman (2001) and Broad and Heckscher (2003). It is instructive to read Bebbington (2004a) and Escobar (2004) in tandem because they provide such different views on anti-globalisation and its livelihood effects in two highland areas of Latin America. Recent years have seen a flowering of studies of place-based 'resistance' movements. Three case studies from Pakistan, Indonesia and the Philippines are provided by Butz (2002), Li (2000, 2001) and Resurreccion (2006); each shows how resistance is never a simple us/them confrontation and how the local is often deeply divided.

Defining resistance

Guha, Ranajit (1998) 'Introduction', in: Ranajit Guha (ed.) *A subaltern studies reader, 1986–1995*, Delhi: Oxford University Press, pp. ix–xxii

Scott, James C. (1985) *Weapons of the weak: everyday forms of peasant resistance*, New Haven, CT: Yale University Press.

Case studies in resistance

Duncan, James S. (2002) 'Embodying colonialism? Domination and resistance in nineteenth-century Ceylonese coffee planatations', *Journal of Historical Geography* 28(3): 317–338.

Isager, Lotte and Ivarsson, Søren (2002) 'Contesting landscapes in Thailand: tree ordination as counter-territorialization', *Critical Asian Studies* 34(3): 395–417.

Ong, Aihwa (1987) *Spirits of resistance and capitalist discipline: factory women in Malaysia*, Albany, NY: State University of New York Press.

Resistance and anti-globalisation

Bebbington, Antony (2004a) 'Movements and modernizations, markets and municipalities: indigenous federations in rural Ecuador', in: Richard Peet and Michael Watts (eds) *Liberation ecologies: environment, development, social movements*, London: Routledge, pp. 394–421.

Broad, Robin and Heckscher, Zahara (2003) 'Before Seattle: the historical roots of the current movement against corporate-led globalisation', *Third World Quarterly* 24: 713–728.

Escobar, Arturo (2004) 'Beyond the Third World: imperial globality, global coloniality and anti-globalisation social movements', *Third World Quarterly* 25(1): 207–230.

Mittelman, James H. (2001) 'Mapping globalisation', *Singapore Journal of Tropical Geography* 22(3): 212–218.

The politics of place and difference

Butz, David (2002) 'Resistance, representation and third space in Shimshal village, northern Pakistan', *Acme* 1: 15–34.

Li, Tania (2000) 'Articulating indigenous identity in Indonesia: resource politics and the tribal slot', *Comparative Studies in Society and History* 42(1): 149–179.

Li, Tania (2001) 'Masyarakat adat, difference, and the limits of recognition in Indonesia's forest zone', *Modern Asian Studies* 35(3): 645–676.

Resurreccion, Bernadette P. (2006) 'Gender, identity and agency in Philippine upland development', *Development and Change* 37(2): 375–400.

9 The structures of the everyday

Introduction

In my view, there is little question that *we* – and I write this assuming that most of the readers of this book will be based in the countries of the Global North – need to know more about the Global South. This is partly for the simple reason it is not possible to consign the countries of the South, viewed for much of the latter half of the twentieth century as the residual 'Third World', to the role of economic and political also-rans. As was noted in the opening chapter, 'emerging' economies now (in 2006) account for more than half of world GDP, measured at purchasing power parity, and China and India are the world's second and fourth largest economies (The Economist 2006b: 3–4). Emerging economies also contribute 43 per cent of world exports. In time, the Security Council of the United Nations may be reformed to more fully represent the countries and peoples of the world and the balance of global political power is shifting in line with the rebalancing of the global economy. The divisions of the world – rich/poor, North/South, Western/non-Western – which we became so comfortable with over the course of the twentieth century have broken down, even if there is a degree of inertia in the constitution of the world's key institutions, and in people's minds.

All that said, it is also true that the great humanitarian challenges of the twenty-first century are to be found in the Global South. The eight Millennium Development Goals (and the underpinning eighteen targets), for example, are almost entirely pitched at the countries of the Global South (Table 9.1). So, while some countries may be becoming relatively richer in terms of their global economic ranking, there is also much to be 'done' reflected in a cavalcade of global summits, appeals, sporting events, concerts and other initiatives. I take the fact that we should be concerned about the Global South as self-evident. It is the majority world. It is becoming increasingly important to 'us' because of its growing economic, environmental and strategic significance at a global level. And there is also a persuasive humanitarian and moral case for concern. The great majority of the world's absolute poor live in the Global South, and almost all those who die before they should are to be found there.

But why should we, from an academic standpoint, be interested – rather than just concerned – with geographies of the Global South? There are two possible answers to this question. First of all there is the possibility that scholarship on the Global South is different in important ways from scholarship on the Global North. That the questions that people ask, and *how* they ask those questions mark academic endeavour in the countries of the South as different. The second possible answer is that it is not only that scholarship is different but that the countries and the peoples of the Global South are, themselves, also different.

Table 9.1 The Millennium Development Goals

Goal	*Target*
1 Eradicate extreme hunger and poverty	• Reduce by half the proportion of people living on less than a dollar a day • Reduce by half the proportion of people who suffer from hunger
2 Achieve universal primary education	• Ensure that all boys and girls complete a full course of primary schooling
3 Promote gender equality and empower women	• Eliminate gender disparity in primary and secondary education preferably by 2005, and at all levels by 2015
4 Reduce child mortality	• Reduce by two-thirds the mortality rate among children under five
5 Improve maternal health	• Reduce by three-quarters the maternal mortality ratio
6 Combat HIV/AIDS, malaria and other diseases	• Halt and begin to reverse the spread of HIV/AIDS • Halt and begin to reverse the incidence of malaria and other major diseases
7 Ensure environmental sustainability	• Integrate the principles of sustainable development into country policies and programmes; reverse loss of environmental resources • Reduce by half the proportion of people without sustainable access to safe drinking water • Achieve significant improvement in lives of at least 100 million slum dwellers, by 2020
8 Develop a global partnership for development with targets for aid, trade and debt relief	• Develop further an open trading and financial system that is rule-based, predictable and non-discriminatory, includes a commitment to good governance, development and poverty reduction – nationally and internationally • Address the least developed countries' special needs. This includes tariff- and quota-free access for their exports; enhanced debt relief for heavily indebted poor countries; cancellation of official bilateral debt; and more generous official development assistance for countries committed to poverty reduction • Address the special needs of landlocked and small island developing states • Deal comprehensively with developing countries' debt problems through national and international measures to make debt sustainable in the long term • In cooperation with the developing countries, develop decent and productive work for youth • In cooperation with pharmaceutical companies, provide access to affordable essential drugs in developing countries • In cooperation with the private sector, make available the benefits of new technologies – especially information and communications technologies

Source: http://www.un.org/millenniumgoals/index.html

Scholarship and the Global South: degrees of difference?

The issue of difference is usually pursued in terms of differences of degree: *more* poverty, *less* education, *worse* health profiles, *lower* levels of technology, and so forth. It we take this view – and it has become the normal view among many geographers – then it provides something of an excuse to ignore or play down the Global South on the basis that the fundamental issues are the same as those in the Global North, even if the context or their intensity may be different. The second possibility is the more interesting, and the more profound: that the geographies of the Global South are different in kind.

In the opening chapter, a brief reference was made to Dipesh Chakrabarty's (2000) book *Provincialising Europe* and his comment that European scholarship is both 'indispensable and inadequate' to the task of illuminating the histories of the non-Western world (Chakrabarty 2000: 16). The same point could equally be made in connection with geographies of the non-Western world: the use of the North as the 'silent referent' for so much of our understanding of the South is inadequate because there is a taken-for-granted attitude that we can begin with, and begin from, the North and use this 'referent' as a guide for the undertaking of geographical analysis in the South.

So, is the Global South different? To be honest, I vacillate on this important question. There are reasons to think that it is different. Historically and politically the Global South is the postcolonial world; culturally and socially, it is the non-Western world; and economically, it is the object of development and the target for investment. These all mark the Global South out as distinctive. Scholars, including geographers, who are area studies specialists build careers analysing and interpreting the special and unique qualities of their chosen countries and regions on the basis that we cannot simply 'export' ideas and experiences from one context and expect them to make sense and be relevant in another. As the discussion of counter-urbanisation in Zambia (see page 140) suggested, we should not be tempted to assume that apparently similar processes being enacted in the Global South have the same underpinning logics to those underway in the Global North.

As an Asianist I am attuned to the uniqueness of place(s). That said, I am also aware of the dangers of over-specification and of failing to look for underlying patterns in the global mosaic. In this era of globalisation, while difference had not been erased, it has meant that connection, interpenetration and interdependency are deeper and stronger than ever before in human history. So – and these are weasel words to be sure – we need to be not only cognisant of the fact of difference at all scales but also attuned to the connections and common processes that link peoples and places. Yeung and Lim (2003: 110) take a slightly different slant on this issue with reference to economic geography, when they highlight what they perceive to be a gap between the *universalising tendencies* of theories coined in the West and based on the experience of the West, and the highly *specific and specifying geographies* of area studies specialists. The middle ground – a ground which takes the experience of the non-Western world into mainstream debates over (economic) geography – is, for them, yet to be populated.

In addition to the question of whether there are differences between the Global North and South, there is the equally important issue of difference within the South. Just as we should be chary of assuming that urban transitions (for example) in Europe and North America will be mirrored across the rest of the world, so we should also resist the temptation to write and think about a universal, or at least a broadly applicable, urban transition for the Global South. There will, in all likelihood, be as much difference between Africa, Asia and Latin America as there will be between those continents and Europe and North America. There could also, of course, be great degrees of difference between countries in each continent.

What, then, of the second question: is scholarship on the Global South different? There are, again, at least two ways to think about this question. To begin with, in terms of the tenor or approach of scholars (and, more particularly, geographers) who work on the Global South. And second, in terms of the details of scholarship. In terms of tenor and approach, research on the South does, I would argue, have certain broad qualities which mark it out as different. Without suggesting that these apply across the board, research into geographies of the Global South, and more particularly the everyday geographies of the Global South, tends to

- be participatory
- emphasise a long-term engagement with places and the people who live there
- adopt a 'grassroots' perspective
- draw heavily on non-geographical work and emphasise cross- or trans-disciplinary links and associations
- focus on empirical issues and be policy-relevant
- eschew heavy theorisation
- be placed-based
- use the household as the object and locus of study
- and acknowledge the role and importance of culture.

To take the last of these as a case in point, while the role of culture in development has recently come back to the fore (see Radcliffe and Laurie 2006), it can be argued that work on the Global South never really 'forgot' the importance of context – including cultural context – in understanding the myriad ways different parts of the world 'work'. This can be seen, for example, in the manner that the concept of social capital (see page 51) has been embraced and applied in the Global South and the Global North. In the Global North, work on social capital led to a redirecting of attention to the role of social organisations and associational life in making societies operate 'efficiently'. In the Global South, social capital became a new term to be applied to structures, networks and organisations which were *already* recognised as central to understanding, for example, how irrigation networks are managed in Bali (Lansing 1991), how marketing systems operate in Ghana (Lyon 2000, 2003; see also Porter and Lyon 2006), and how, in wider terms, capitalism functions in Asia (Yeung and Lim 2003; Yeung 2004). Contemporary geographical debates are often highly nuanced in their treatment of politics, identity, consumption practices, and so forth. And yet they are also sometimes surprisingly blind in their assumption that these nuanced perspectives apply cross-culturally.

There are also some more specific attributes which sets research in/on the South apart from research in/on the North. These include, for example, a concern for and interest in: community strategies and grounded, bottom-up views; environmental risk and livelihoods; local-level transitions; and indigenous knowledges. It is also true that, gradually, almost imperceptibly, there has arisen some reverse learning where methods and approaches pioneered in the South have begun to find favour and popularity in the North. Korf and Oughton (2006), for example, in an article entitled 'Rethinking the European countryside – can we learn from the South?' propose that:

> in view of the much longer tradition in bottom-up approaches, interdisciplinary poverty studies and broader analytical concepts in development studies in the South, it may be useful to explore how appropriate such methods may be in studying and addressing similar issues in the North.
>
> (Korf and Oughton 2006: 278–279)

Illustration 9.1 A Northern scholar extracting information from people in the South (in this instance, Laos).

Using participatory poverty assessments (see page 76) and the sustainable livelihoods approach (see page 29) as their 'Southern' methodologies, they show that while there are problems connected with applying such approaches uncritically in the North (of course!), there is much that can be learnt from a greater and more balanced exchange of ideas and experiences.

People reading this book will likely have become aware that while it is about the Global South, it is written by someone from the Global North. The great majority of the work referred to in the book is also authored by Northern scholars based, in large measure, in Northern universities. We have grown used to this, and it is regarded as normal. But I sometimes wonder how I would feel if Durham became the site of intensive research by a caravan of scholars from Asia who were forced to use interpreters because they didn't know the language, and who published their results, interpretations and views in journals that I couldn't read and, if I could, couldn't afford to purchase (Illustration 9.1). This gives just an inkling of the degree to which a certain structure, balance and flow of scholarship and scholarly endeavour has come to be regarded as normal.

What this book doesn't do

This book should not been seen as comprehensive in terms of coverage, whether geographically, thematically or in terms of approach. There are highly important questions and issues which are not addressed, or at least not head on. Perhaps the greatest of these relates to global poverty and the concentration of the world's absolute poor in the countries of the Global South: Why is the Global South still – generally speaking – the poor world?

A structural approach to the geography of the Global South would have highlighted, for instance, the unfair trade policies of the rich world, the actions of the IMF, structural adjustment and, no doubt, the effects of globalisation. None of these important issues has been discussed in detail in the book. But I hope that the discussion has gone some way to showing why such macro-level perspectives are, on their own, insufficient in building an understanding of why individuals, households and communities are poor. The evidence of the effects of globalisation on poverty is uncertain and general statements, one way or the other, are unconvincing (see Bardhan 2006). It is only by drilling down to examine how the different elements of globalisation (or structural adjustment, or land reform, or new technologies for rice cultivation) penetrate national economies, regional spaces, and household contexts that we can begin to reach for an answer. So, for instance, in Zambia liberalisation policies have not led to the expected increases in farm productivity; in Mexico, price incentives have had limited effects on output (examples quoted in Bardhan 2006: 1395); while in Asia, roads do not always deliver the benefits expected and may actually harm some people. In the Zambia case the identified reason was that farmers lacked draught animals and farm implements to extend production and increase output in response to liberalisation; in the case of Mexico, it was because rural households could not access capital to take advantage of price incentives; and the study of the effects of roads on the rural poor in Asia demonstrated how road improvements are filtered and sifted according to prevailing social structures, cultural norms, and the distribution of productive assets (Table 9.2) (Hettige 2006: 6).

In other words, the policies themselves leave open a wide range of possible outcomes, for countries, sectors, activities and, ultimately, for households and individuals (see Box 9.1). If economic growth is driven by technological progress, and if technological progress rewards those people with skills and education, then those without these capabilities may well suffer from declining returns to their work, both relatively and absolutely (Thorbecke and Nissanke 2006: 1334). Geographers have become rather adept at highlighting the world of flows, nodes and linkages that globalisation has created; they have been rather less skilled, however, at unpicking the localised, place-based dynamics which both create and are a creation of these flows (Bebbington 2003). To return to a theme addressed in the opening chapter, it is at the

Table 9.2 Do roads deliver, what and to whom?

Category	*Effects of improved roads/transport*
Women	Women are less likely to be able to take advantage of improving transport facilities, even when cost is not an issue, because they face social barriers to mobility such as the stigma of riding a bicycle or travelling alone outside their community.
Women and men	Women and men have different transport needs. Women's needs tend to be for frequent, local journeys; men for less frequent, longer trips. Women's needs are directed at meeting household consumption requirements; men's for income generation and production.
Rich and poor	Richer families have the time to travel, the products to sell, and the money to purchase goods. The poor are short of time and money and better roads often do not increase incomes because they have nothing to sell.
Very poor	The very poor usually walk and 'inhabit a localized walking world' (Hettige 2006: 18); roads deliver little for this marginal and marginalised group.

Source: information extracted from Hettige 2006

BOX 9.1 Scale, history, social relations and livelihoods in Syria

In 1958 the government of Syria implemented a land reform programme to redistribute land and create a small holder class of owner occupiers. This was partly driven by equity concerns in the countryside although another objective was to channel private wealth into industrial investment and away from land accumulation. In the half century since this land reform, holdings have declined through inheritance and land division so that today, in many areas, land holdings are sub-livelihood in extent despite attempts to drive up yields (and therefore output) through agricultural intensification. The result, today, is that many households have been forced to diversify by seeking off-farm work and, often, this has involved migration. Unlike the Indonesian, Thai and Indian case studies presented in other chapters of this book, these migrants from the Syrian countryside are largely male. Women stay at home.

Abdelali-Martini et al. (2003) discuss this process in the Syrian countryside drawing on fieldwork undertaken in Aleppo and Idleb in north-west Syria. They note that migrants are usually sons; male heads of household stay at home. Because of the loss of male labour, women are forced to take on more agricultural work, leading to a feminisation of agricultural labour. This may include work on the household's land and, in the case of those with very small holdings, the farms of others in the local area as well. But because either the male head of household or the eldest son remain at home, it does not lead to a feminisation of farm management or of agriculture *in toto*. Women do not become de facto household heads. They may work more, but have little more authority.

In their article, Abdelali-Martini et al. (2003) demonstrate the links between different scales and different contexts:

- A national land reform programme, partially motivated by a desire to fund and support industrial development, reworks land holding in the countryside.
- Progressively, over half a century, the programme sets the context for a squeeze on rural livelihoods which is met by livelihood diversification.
- Off-farm (urban and international) work and the opportunities to access this work expand with regional development informed by globalisation.
- Cultural norms create gender-selective migration streams, leaving women at home while men migrate.
- Gender roles, nonetheless, come under pressure as the necessity to work the land forces a feminisation of agricultural labour.
- But this does not lead to a deeper change in gender relations because of the continued presence of a male household head.

Source: Abdelali-Martini et al. 2003

interface between structures and agencies, between institutions and actors, between global flows and local dynamics, where we need greater analytical purchase. And it is at this level, at the local and the everyday, where *d*evelopment and *D*evelopment (i.e. the Development project) are mediated.

A recurrent theme of this book has been to highlight the gap between the aggregate and the national/global and the particular, the personal and the local. We can scrutinise national policies in states with apparently effective bureaucracies, analyse aggregate statistics collected by well-functioning statistical agencies, and wonder at the efficacy of international organisations and the influence and power of global structures and processes. Yet, at the local level and in terms of understanding people's lives, their actions and the choices that they make, all of these are found wanting in terms of their explanatory bite. Policies are rarely transmitted in the form and manner intended. Even more rarely are they 'implemented' in such a manner. Agencies are compromised and compromise at every turn. Aggregate statistics and broad trajectories of change are shown to be generalised to a fault.

The reasons for this 'gap' are several. To begin with the most obvious: the world is complex and variable. This is the underpinning tyranny of studies which take the local and the personal as their starting points. Second, economics and politics – often seen as occupying the commanding heights of the social sciences – are interpenetrated by the social and the cultural. And the social and the cultural arise from local contexts. Victoria Beard's articles (1999, 2002, 2003) on community and 'covert' planning in Indonesia provide an example of the power of the local to defy the machinations and managerialist tendencies of the state. In each article she asks herself the question, 'How do ordinary people undertake acts of social transformation in authoritarian contexts where the threat of physical violence is present?' In response, she argues that while the Indonesian state, until recently, was authoritarian and undertook planning in a top-down and hierarchical manner this never obliterated opportunities for community-organised and directed planning which operated beyond the sight and the scope of the state. In line with the discussion in Chapter 8, the ability of ordinary people to do this in Indonesia relies on covert action: 'Local people are constantly maneuvering, navigating, and creating spaces that challenge existing power relations in subtle and incremental ways' (Beard 1999: 128). In some instances this may not just rework local spaces but resonate up the lines of authority and communication and force change at the centre (as in the case of Vietnam discussed in Chapter 8).

In writing this book, I have also tried to put some distance between the geographies that are described and interpreted, and 'development' (as in the development project). It is reasonable to question, however, whether it is ever possible to understand the Global South distanced from the literatures and ideas about development and the policies, programmes, agencies and institutions which pursue and justify the development industry. While the discussion has, however, largely avoided discussion of *D*evelopment, *d*evelopment has figured strongly. The focus on everyday geographies has permitted the development project to be reframed and contextualised within the context of the personal geographies of development (Box 9.2).

Much has been written on the search for 'alternatives to development' in the hope that these might offer a future for the Global South which does not rely, intellectually or practically, on the Global North. When, however, development is stripped of its grand project clothing and is seen, instead, as a locally embedded process of change then its contingency becomes all too clear. Modernity is shot through with elements of pre-development. The development project has not so much multiple trajectories, as multiple forms which may, or may not, result in what appear to be similar trajectories. And while *D*evelopment may be all too obvious in the tangible efforts and grand pronouncements of governments, agencies and NGOs, behind and beneath all this effort and discourse is *d*evelopment – covert, unsung, often a whisper rather than a roar but nevertheless highly influential in forging the everyday geographies with which this book has been concerned. In other words, it is important to *look*

BOX 9.2 The Millennium Development Goals: localising the MDGs

The Millennium Development Goals, adopted by the General Assembly of the United Nations on 18 September 2000, represent the grandest of all development initiatives. The eight goals, to be achieved by 2015, provide a handy wish list signed-up to by nearly 190 countries. Yet, virtually from the start, there were scholars who questioned whether this apparently self-evidently desirable set of goals really targeted the heart of the development challenge in the Global South. Moreover, this questioning often focused on the need to contextualise the goals and the associated targets in terms of the local, the personal and the everyday.

James (2006) directs his attention at what he perceives to be a failure to distinguish between means and ends, or between *actual* achievements (ends) and *potential* achievements. Some of the MDG targets are ends, manifested and measurable at the level of the individual. This applies to the health-related targets under goals 4, 5 and 6 (see Table 9.1). But many of the other goals are, in James' view, means rather than ends. So, for example, he draws a distinction between completing primary school (Target 3 under Goal 2) and the acquisition of basic literacy and numeracy. The former (primary school education) may lead to the latter (literacy and numeracy), but if schooling is inadequate, as it so often is in the poorest countries and in poorer places, then this may not be achieved. In other words, the meeting of a target – universal primary level education – may not deliver the desired end of an adequate education for the modern world.

Satterthwaite's (2003) two core concerns regarding the MDGs again direct our attention at local context. To begin with, he has doubts about the aggregate statistics on which achievement of the targets is based, and particularly those for the very poorest countries. He writes of 'nonsense' statistics, such as the levels of urban poverty and urban service provision in Africa. 'Anyone with any knowledge of Nairobi', he writes, 'would be astonished to see that only 1.2 per cent of Kenya's urban population was considered "poor" in 1998, or that only 4 per cent of its urban population lacked sanitation' (Satterthwaite 2003: 184). A second concern relates to the use of measures which are devised by experts, informed by international agencies, assessed by governments, and applied irrespective of local context. The core MDG poverty target, for example, is income-based. Other forms of deprivation linked to social exclusion, political marginality and cultural rights, which are often local in character and difficult to scale up, are ignored and the inequalities in power that are so often the root cause of poverty, overlooked (Satterthwaite 2003: 182).

into the trajectories of development and change evident across the Global South. It is what the trajectories contain which, I suggest, is more important in defining difference than the trajectories themselves. Most societies are modernising in ways that would be familiar, I sense, to all of us; but within these modernising tendencies, this modernising teleology, is a great degree of difference. And it is in this context and sense that it is important to explore everyday geographies.

The scale debate revisited: the places of the Global South

The question of whether it is possible, or desirable, to think and write in scalar terms was discussed in Chapter 1. It was argued that scale and the role of place remain powerful and significant contextual axes. Like Mansfield (2005) and Jonas (2006), I find much of the rescaling and descaling arguments unhelpful in understanding the everyday geographies of the Global South. They are theoretically rich but, often, empirically problematic: 'methodologically, geographical research has to start some*where*' (Jonas 2006: 403, emphasis in original). Those who would wish to develop a human geography without scale (e.g. Marston et al. 2005) maintain that by using scale as an explanatory scaffold, it inevitably becomes hierarchical, and privileges a particular way of seeing (whether local or global). Even when scales are nested there is 'a foundational hierarchy – a verticality that structures the nesting so central to the concept of scale, and with it, the local-to-global paradigm' (Marston et al. 2005: 419).

In making the case for a geography of the everyday which is also a geography of places and people in places, I am not suggesting that analysis should be tied unremittingly to localities. The discussion of migration and mobility in Chapter 6 should have emphasised that people across the Global South are moving further and with greater frequency than ever before. Livelihoods are becoming delocalised, identities are becoming trans-local, and people are living and making a living across spaces. The discussions in Chapters 7 and 8, meanwhile, have shown how local political acts are intimately tied up with other political agents and agencies based in other places and at other scales. It is just that the approach that this book has taken has been to start with the local (and the personal), and then to trace these geographies to other sites, scales and contexts. This does, inevitably, provide a particular and a partial view.

What this book has tried to do, perhaps above all else, is to emphasise the Global South and its myriad forms as places that are important in and of themselves. The countries and peoples of the Global South cannot – and should not – be reduced to bit players in a global economic, cultural and political project, dim reflections of what has or is going on in the Global North. Scholars, for many years, have been at pains to emphasise local context, to stress that progress is not uni-linear, to nod in the direction of the importance of place, and to press for greater substantive engagement with the Global South. But this rhetorical engagement with the places of the Global South has often not been matched by a commitment to research on people in places:

> The disregard of place in the social and human sciences is the most puzzling since . . . it is our inevitable immersion in place, and not the absoluteness of space, that has ontological priority in the generation of life and the real. . . . We are, in short, placelings.
> (Escobar 2001: 143)[1]

Through the chapters of this book, the role of place has arisen in many contexts, with reference, for instance, to identities, resistance, cultural politics, livelihoods, and mobility and remittance landscapes. This is not to suggest that peoples and places are somehow unconnected to the extra-local. The point is that we need to see today's network society embedded in places, rather than networks erasing places. It is not that global capital displaces local livelihood and economic forms; these local forms become part of global capitalism but do so in particular ways. So it is the interrelationships and the scalar-links that exist in places which make places special. It also means that fieldwork (broadly defined) must lie at the heart of an understanding of the everyday geographies of the Global South. As Mallon says with reference to subaltern strategies in Latin America:

> In my experience, it is the process itself [of research] that keeps us honest: getting one's hands dirty in the archival dust, one's shoes encrusted in the mud of fieldwork; confronting the surprises, ambivalences, and unfair choices of daily life, both our own and those of our subjects.
>
> (Mallon 1994: 1507)

As will have been clear from the foregoing chapters, this book – through its use of grounded case studies and examples – has made a case for the driving importance of place in understanding the everyday geographies of the Global South. In calling for the '"repatriation" of place in anthropology', Arturo Escobar (2001: 142) is echoing the views of some geographers who, too, have called for a re-engagement with the specificities of the multiple geographies of the Global South.

Perhaps part of the reluctance among geographers to engage in place-based scholarship lies with the implicit links that are made to earlier place-based tendencies – in particular, regional geography and area studies. Scholars who became area studies specialists found themselves labelled as 'ists': Asianists, Africanists, Latin Americanists . . . and, as a result, their careers sullied. As Potter has observed, 'those who work outside the Euro-North American orbit are excluded, or at least maginalized, from the specialisms that see themselves as making up the *core* of the discipline of *Geography*. Quite simply, they are regarded as "ists"' (Potter 2001: 423, emphases in original). At the same time, those geographers who have made the Global South their stage of study – rather than any one part of it – tend to be categorised as 'development geographers'. Once again, this tendency also places their work outside the mainstream of geography with the result, I would say, that it comes to be regarded not just as marginal, but also as inferior.

Mrs Chandaeng: a reprise

The opening lines of Chapter 1 introduced Mrs Chandaeng, the widowed and landless mother of six living in a village close to the Mekong. The Lao People's Democratic Republic, the country where she happens to live, has become the object of development. The World Bank, International Monetary Fund, Asian Development Bank and a myriad of bilateral aid agencies, international NGOs and expert advisers populate the capital, Vientiane. These agents of the global international development industry have been joined in recent years by a growing army of foreign investors, tempted by the profits they think they can generate in this resource-rich land. Directed by the New Economic Mechanism, the country and its economy are being 'reformed' to bring them in line with the neo-liberal Washington consensus, instilling fiscal discipline, freeing-up the financial sector, liberalising trade, privatising state-owned enterprises, deregulating the economy, enticing foreign investment, and reforming the tax system. According to one book on foreign investment, the '[Lao] government's reform campaign has been fully integrated, monitored, and analyzed' (Sunshine 1995: 2). Even if that were so, however, Laos' reform programme provides only a modicum of illumination by which to read and understand Mrs Chandaeng's life story.

Her life story reveals a combination of structure, agency, serendipity and misfortune. It also links culture, history, economy and environment. Mrs Chandaeng's early life in the province of Xieng Khouang was marked by the Vietnam War and the bombing campaign which devastated the area, and the final victory of the communist Pathet Lao in 1975 which effectively cut the provinces of Laos off from one another, and the country from much of the wider world. The sudden death of her husband in 1988 was a stroke of misfortune which, coupled with a falling-out with her brother-in-law, forced Mrs Chandaeng to leave home and search for a new life. She settled in Ban Sawai, but was hardly well placed to build a sustainable livelihood with six children, no land and little education. But the porosity of the international border with Thailand, an (almost) common language, and the economic dynamism and growing wealth of Thailand provided a context where her children, propelled by lack of opportunities at home and possessing the courage to look over the horizon, could seek out alternative jobs outside farming and beyond the local area. The opportunity for her three daughters to leave home and work in Bangkok and her son to work on a shrimp farm was made possible by a series of economic and political reforms in Laos, and the dynamism of a neighbouring country – Thailand. But it was also permitted by another series of changes in cultural and social norms that were being contested and gradually renegotiated at the household and community levels right across Laos. One influence, no doubt, was Thai television, which is beamed across the Mekong and provides an all-too-tempting view of modern lives in another land where young women can become *than samai* and up-to-date and where money is easy to come by. Such work also, though, provides the means by which daughters can remain dutiful through channelling funds back to their natal villages to sustain and perhaps transform the lives of those who have remained at home. It is neither easy nor desirable to package the experience of Mrs Chandaeng as either local, trans-local, glocal or global. It is all and none of these. What we do see, however, in her and her family's lives is the presence and operation, simultaneously, of networks, places, scales, flows, nodes, structures and agencies. Mrs Chandaeng – her everyday geographies – provide not a narrow and restrictive vision but one which opens up a window of understanding onto geographies above and beyond the local and the everyday.

Further reading

There is a body of work which makes a case for a redirecting and rebalancing of attention away from the North and the West. Yeung and Lim (2003) and Yeung (2004) highlight the need for economic geography to develop a theoretical and conceptual base which takes other places and contexts seriously, while Korf and Oughton (2006) apply some methodological approaches developed in the South to Northern contexts as part of an effort at 'reverse learning'. Escobar (2001) and Bebbington (2003), in their different ways, both make sustained cases for the need for more place-based geographies, while the articles by Marston et al. (2005) and Jonas (2006) provide very different perspectives on scale. Beard's three articles (1999, 2002, 2003) on 'covert' planning in Indonesia, together constitute an effective case study of how national planning becomes localised and how local people can manipulate even authoritarian and hierarchical systems.

Southern geographies?

Korf, Benedikt and Oughton, Elizabeth (2006) 'Rethinking the European countryside – can we learn from the South?', *Journal of Rural Studies* 22: 278–289.

Yeung, Henry W-C. (2004) *Chinese capitalism in a global era: towards hybrid capitalism*, London: Routledge.

Yeung, Henry W-C. and Lim, George C.S. (2003) 'Theorizing economic geographies of Asia', *Economic Geography* 79(2): 107–128.

Defending and valorising place

Bebbington, Anthony (2003) 'Global networks and local developments: agendas for development geography', *Tijdschrift voor Economische en Sociale Geografie* 94(3): 297–309.
Escobar, Arturo (2001) 'Culture sits in places: reflections on globalism and subaltern strategies of localization', *Political Geography* 20: 139–174.

Scale debates

Jonas, Andrew E.G. (2006) 'Pro scale: further reflections on the "scale debate" in human geography', *Transactions of the Institute of British Geographers* 31: 399–406.
Marston, Sallie A., Jones III, John Paul and Woodward, Keith (2005) 'Human geography without scale', *Transactions of the Institute of British Geographers* 30: 416–432.

Structures and agencies

Beard, Victoria (1999) 'Navigating and creating spaces: an Indonesian community's struggle for land tenure', *Plurimondi* 1(2): 127–145.
Beard, Victoria (2002) 'Covert planning for social transformation in Indonesia', *Journal of Planning Education and Research* 22: 15–25.
Beard, Victoria (2003) 'Learning radical planning: the power of collective action', *Planning Theory* 2(1): 13–35.

Notes

1 What's with the everyday?

1 Lord May noted a similar imbalance when, in his anniversary address to the Royal Society in 2005, he noted that the proportion of articles in the four leading international medical journals devoted to health issues in developing countries was just 15 per cent in January 2002 and 12 per cent in January 2003. A large number of these were related to a single health concern – HIV/AIDS (May 2005: 15). Research and development funding in 2001–2002 by disease per disability-adjusted life year (DALY) amounted to $63.45 for cardiovascular disease (a disease of affluence and rich countries) and $6.20 for malaria (a disease of poor countries and impoverished people) (The Economist 2006a: 70).

2 Yeung and Lim (2003) in their review of economic geography also comment on the tendency of the subdiscipline to 'address theoretical and empirical issues specific to only a handful of advanced industrialized economies' (2003: 108). The scholars associated with the key theoretical advances in economic geography since the 1960s have all come from Anglo-American countries and most do their empirical work in advanced economies.

3 I appreciate that not everyone likes the term 'Global South'. They find it ugly and ponderous. It is used here because the alternatives – 'developing world', 'Third World', 'poor world' and so on (see Box 1.1) – are viewed as even more problematic. Before long, no doubt, another term will replace it.

4 To date most of my fieldwork has been undertaken in Thailand, Vietnam, Cambodia, Laos and Indonesia.

5

> . . . the eco-localist critique of globalization does not depend only on the impact of trade treaties or international economic institutions on the environment. Instead, it is based on the impact of long supply chains, the greater separation of production and consumption, cheaper prices and consumerism. Globalization harms the environment, from this perspective, both by increasing throughput per unit consumed and by increasing overall material consumption. . . . ecolocalists conclude that the only way to create economic sustainability is to (re)localize the economy.
>
> (Curtis 2004: 98)

2 Structures and agencies

1 Archer calls the subsuming of structure into agency 'upwards conflation' and the opposite (agency into structure) 'downwards conflation' (Archer 1995: 33).

2 Although this is the source and the date usually given to mark the beginning of the SL perspective, Chambers also used the term sustainable livelihoods in an earlier IDS discussion paper published in 1988, with reference to 'sustainable livelihood security'. The philosophy also informs much of the very widely read 'Brundtland report' of the World Commission on Environment and Development (1987).

3 This is explored in greater depth in Chapter 3.

4 Shea nuts, when crushed and processed, yield a 'butter' sometimes known as 'women's gold'. This is used in local foods but has also become an important and valuable commodity, particularly for women and the poor, as it is used in chocolate and cosmetics production. The shea tree is found in

semi-arid areas of the Sahel region of West Africa. (See Harsch, Ernest (2001) 'Making trade work for poor women: villagers in Burkina Faso discover an opening in the global market', *Africa Recovery* 15(4): 6, downloaded from http://www.un.org/ecosocdev/geninfo/afrec/vol15no4/154shea.htm.)

5 This is another reason why this book about the Global South is not about development.
6 Working papers, methodological statements, and the rationale of the project can be accessed from http://www.welldev.org.uk/research/research.htm.
7 This is a summary based on Leach et al. (1999).
8 These two terms are taken from a study by Murray (2002).
9 To paraphrase the nineteenth century French economist Leon Walras, it can be argued that we cannot explain anything until we have explained everything (The Economist 2006c: 75). All is linked; all is relevant; and everything has potential utility.

3 Lifestyles and life courses

1 Sen's (1990a) article proved to be particularly controversial (see Sen 2003; Oster 2005). However, while the sheer size of the figure (100 million) has been disputed and the application of the reasoning to some countries questioned the fact that there is a large imbalance is generally accepted.
2 In a later article, Sen (2003) explored the widening gap in the number of male versus female foetuses born in East Asia. The normal ratio is 95 girls born to every 100 boys. The figure for Singapore and Taiwan is 92; for South Korea, 88; and for China, 86 (Sen 2003). This has been largely associated with the widening availability of sex-specific abortion.
3 While Putnam's work on social capital may be more familiar, Bourdieu can reasonably claim to be the 'father' of social capital.
4 For critical accounts see Harriss and Renzio (1997), Putzel (1997) and Fine (2002). For a response to these, see Bebbington (2002, 2004b).
5 In their article, Bebbington and Perreault (1999) do not highlight the relations of trust, reciprocity and exchange or the common rules, norms and sanctions that some other studies have emphasised as important elements in social capital (see Table 3.2). However, one can assume that the formal establishment of 'communities' has created a context in which these have been developed or bolstered.
6 While the term 'life cycle' is better known than 'life course', the latter tends to be preferred by scholars because it alludes to a series of transitions or transformations, rather than a 'cycle' of return and re-return. When adolescents become adults they never return to being adolescents.
7 Arnold (2005) sets out to show how the British in early colonial India (late eighteenth century) tried to deploy a strategy of 'improvement' to deal with the poverty and destitution they saw in the countryside, whether through the 'anglicisation' (reforming rural relations along English lines) or 'tropicalisation' (introducing systems of estate crop management akin to those in the West Indies and Brazil) of agriculture in the subcontinent. He concludes, however, that the 'quest for improvement remained largely as it had begun, a European movement, and failed to generate among Indians the necessary enthusiasm and resources for substantial agrarian change' (Arnold 2005: 523).
8 The essence of the King's 'New Theory' was incorporated into Thailand's Ninth Five-Year Economic and Social Development Plan (2002–2006) which has, at its core, the notion of the 'sufficiency economy' (http://www.nesdb.go.th/plan/data/SumPlan9Eng/menu.html).
9 Ferguson's interpretation of modernisation/urbanisation processes in Zambia's Copperbelt has been debated at length in the pages of the *Journal of Southern African Studies*: see Ferguson (1990a, 1990b, 1994) and Macmillan (1993, 1996).

4 Making a living in the Global South

1 See Koizumi (1992) for a discussion of the payment of tax in cash in mid-nineteenth century north-east Thailand.
2 Which may be one reason, Cleary (1996: 320) hazards, why indigenous involvement in the colonial mineral and plantation sectors was so unenthusiastic.
3 100 taka = approx US$1.50.
4 Chance and the unknown is a theme of Ecclesiastes:

> I saw that under the sun the race is not to the swift, nor the battle to the strong, nor bread to the wise, nor riches to the intelligent, nor favour to the skilful; but time and chance happen to them all. For no one can anticipate the time of disaster.
>
> (Ecclesiastes 9:11–12)

5 The detailed methodology employed is set out in World Bank (1999) *Methodology guide: consultations with the poor*, Poverty Group, PREM, Washington, DC. This can be downloaded from http://www1.worldbank.org/prem/poverty/voices/reports/method/method.pdf.
6 In 2015, it is estimated that 58 per cent of Sub-Saharan Africa's population will continue to reside in rural areas, and 66 per cent of the population of South Asia.
7 Although indigenes will often pay less for a given parcel of land than non-indigenous settlers.
8 Demographic data for Asia can be accessed from http://www.adb.org/Documents/Books/Key_Indicators/2005/pdf/rt06.pdf. In China, strict controls on rural–urban migration, policed through the hukou system, kept the urban population as a proportion of the total relatively constant over the 25 years between 1960 and the mid-1980s, at around 20 per cent (see page 123).
9 Ten decimals = 1 Katha; 10 Katha = 1 acre. One decimal = 0.004 ha; 50 decimals = 0.2 ha.
10 The distinction here is between those individuals or households who have left the countryside and permanently relocated to Dhaka and those men who have migrated to Dhaka to secure work and left their families in the rural village of origin.
11 Surprisingly, the Bangladesh PPA does not discuss or mention the particular plight of street children although it does note that there are 6.3 million children under the age of 14 in the work-force.
12 See Lynch (2005) for a general discussion of rural–urban interactions in the developing world and also the articles in the themed issues of *Environment and Urbanization* (1998, 'Beyond the rural–urban divide', and 2003, 'Rural–urban transformations').
13 For examples of studies from Bangladesh see DFID (2004), Garrett and Chowdhury (2004) and Hossain (2004).
14 This distinction is not a sharp one. Some coping strategies (family splitting) rely on certain adaptive transformations (migration).
15 Kritzinger et al. (2004: 31–32) outline how poor farm labourers cope with low and unstable incomes connected with the casualisation of work in post-Apartheid South Africa. They note: buying more maize and less meat; purchasing food in bulk; buying food on credit; baking rather than buying bread; borrowing money from relatives, neighbours or employers; requesting to be paid in advance; saving money during good times; selling cigarettes, brooms, doilies and other products; and taking on domestic work.
16 This was also identified as the second most prevalent shock for medium well-being households.
17 In other words, economic 'growth' has been –3.9 per cent per annum between 1990 and 2003 (http://siteresources.worldbank.org/INTWDR2005/Resources/wdr2005_selected_indicators.pdf). In 2003 the DRC's Human Development Index placed it 167th out of 177 countries (http://hdr.undp.org/reports/global/2005/pdf/HDR05_HDI.pdf). The country's human development index was lower in 2003 (at 0.385) than it had been in 1975 (when it was 0.414).
18 This association is disputed (see Ives and Messerli 1989).
19 The previous serious flood was in 1988.
20 The full report can be downloaded from http://www1.worldbank.org/prem/poverty/voices/reports/national/banglade.pdf.
21 The following discussion draws on fieldwork undertaken in Thailand in July 1995, funded by the US National Science Foundation. Rigg et al. (2006) provide a summary of the research.
22 See Sustainable Livelihoods in Southern Africa (SLSA 2003) for a discussion of reform and livelihoods with reference to southern Africa.
23 As, indeed, is this one.

5 Living with modernity

1 Clearly a country of 1.3 billion people and with a land area of almost 10 million square kilometres will show a degree of variability. In particular, the fast-industrialising coastal zone can be contrasted with a relatively stagnant interior. Growing levels of human mobility, however, provide the means by which ideas, experiences, commodities and money can flow between and link such areas (see Chapter 6).

2 We should not assume that this is necessarily a recent change. Bruner also worked among the Batak of North Sumatra and believes that the value attached to education was in place before the Japanese occupied the Netherlands Indies in 1942: 'Education was the "golden plough", the means of escaping from the drudgery of work in the rice field and the monotony of village life' (Bruner 1961: 511).
3 The term 'immiserising growth' was coined by the economist Jagdish Bhagwati, who used it in a rather different context. He applied it to situations where the expansion of production of an export commodity could lead to global over-supply, depressing values and therefore reducing returns by shifting the terms of trade against the commodity in question (Bhagwati 1958). Today, however, it is used more commonly to (counter-intuitively) mean economic expansion that leads to a fall in overall living standards (see http://www.soc.duke.edu/sloan_2004/Papers/Memos/Kaplinsky_immiserising%20growth_25June04.pdf#search=%22immiserising%20growth%22).
4 In his 1990 book on Mexico's 'distorted development', Barkin writes how farmers have been 'wrenched from their local communities and regional cultures into a new national policy and increasingly subjugated to the designs of an international market' (1990: 3).
5 The authors do note that this finding contrasts with that of Katz, who worked in the same area and found that 70 per cent of married women were excluded from land-use decisions (see Katz 1995: 332).
6 Supermarkets, of course, can – and do – also enter into contractual agreements with producers.
7 For general defences of FDI-driven export-led growth see Bhagwati (2004) and Wolf (2004).
8 See Elmhirst (1995a, 1995b, 1996, 1997, 1998a, 1998b, 1998c, 1999, 2000, 2002, 2004).
9 Like the Lampungese, Sundanese women (and unlike most other women in East and Central Java) were also traditionally confined to the home and were not expected nor permitted to work outside the household. These female factory workers, however, did not leave home long term like Elmhirst's workers but commuted from rural villages to their places of work.
10 This provides another compelling reason why it is necessary to avoid the individualisation of analysis – individuals live in families or households and it is only in terms of this wider context that individual choices, actions and activities can be understood (see page 44).

6 Living on the move

1 Writing of Nepal, Gill (2003: 21) says: 'Good transport links emerge as a key to the creation of new livelihood opportunities in the rural areas'.
2 The discussion in Chapter 5 of female migrants from Lampung in Indonesia is relevant here (see page 108).
3 This study is part of a larger project which maps out four different migration streams in West Bengal, each with its own impetus, logic and outcome. See Rogaly and Rafique's (2003) for an account of seasonal migration and vulnerability in Murshidabad district.
4 The discussion in this sub-section on migration and livelihoods in India is taken from Deshingkar and Start (2003).
5 See, for example, de Haan's (2004) case study of Calcutta's labour migrants.
6 It is worth noting how far this traditional Melanesian world view resonates with the new mobilities paradigm outlined for modern Europeans and North Americans by Sheller and Urry (2006) and discussed at the beginning of the chapter.
7 For a general account of mobility among Sahelian peoples and, particularly, among the Fulbe, see de Bruijn and van Dijk (2003).
8 While it has become commonplace to observe that the poor are less likely – because they are less able – to migrate, there are many studies that find the reverse. One study of Ghana and Egypt, for example, found that the poor and very poor in Ghana, and the poor in Egypt, were more likely to engage in international migration than the non-poor (Sabates-Wheeler et al. 2005).
9 Compare this with the discussion of Lampungese female migrants in Chapter 5 (see page 108).
10 This also resonates with the example from China noted in the previous section.
11 The theme of migrant identities and the politics of difference is returned to in Chapter 8 (see page 178).
12 Around 1.2 million are thought to be legal, registered migrants. See *Migration News* 11(3), July 2004, downloaded from http://migration.ucdavis.edu/mn/more.php?id=3033_0_3_0) and *Migration News* 12(4), October 2005, downloaded from http://migration.ucdavis.edu/mn/more.php?id=3145_0_3_0.

13 It has also been argued that transnational location can harden a sense of national identity. Kong (1999) makes this case in the context of Singaporean nationals living in China where, she argues, a sense of 'Singapore-ness' has been accentuated due to their expatriate experiences. National identity, in other words, is not dependent on physical presence in that national territory – sometimes, quite the reverse.

14 But also see the observation on page 113 regarding the distinction between *gender relations* and *gender roles*.

15 North Subang has been extensively studied: by Hayami and Kikuchi (1981) in the 1970s; by White and Wiradi (1989) in 1981; by Breman (1995) in 1990; by Pincus (1996) in 1990–1991; and by Breman (Breman and Wiradi 2002) in 1998–1999.

16 This brings to mind Tancredi's comment in *Il Gattopardo* (*The Leopard*) by Giuseppe Tomasi di Lempedusa: 'For things to remain the same, everything must change'.

17 While the general picture may be that migrants from rural areas maintain links with their villages of origin, and a commitment to those villages, this does not necessarily mean that they intend to return. In a survey of 997 rural migrants to three urban areas in Afghanistan (Kabul, Herat and Jalalabad) undertaken in 2004–2005, while nearly 80 per cent sent remittances to their villages of origin, just 21 per cent planned to return there. More than two-thirds had either settled or were intending to settle permanently in the city (Opel 2005).

18 While HIV-AIDS may account for some of this decline in Zambia's level of urbanisation (HIV-AIDS rates are highest in urban areas) it is out-migration from urban areas which is most important (see Potts 2005: 594–597).

19 See Potts (1995) and Potts (2005: 601) for reference to relevant literature.

20 De Haan (2002: 128) argues this in terms of the historical experience of Saran district in West Bihar.

7 Governing the everyday

1 To quote the section in full:

> Everyday politics occurs where people live and work and involves people embracing, adjusting to, or contesting norms and rules regarding authority over, production of, or allocation of resources. It involves quiet, mundane, and subtle expressions and acts that indirectly and for the most part privately endorse, modify, or resist prevailing procedures, rules, regulations, or order. Everyday politics involves little or no organization. It features the activities of individuals and small groups as they make a living, raise their families, wrestle with daily problems, and deal with others like themselves who are relatively powerless and with powerful superiors and others.
> (Kerkvliet 2005: 22)

2 Drawing on the work of Foucault, governmentality is the attempt to create or constitute governable subjects or the art of rule.

3 See Box 7.2 for a discussion of the Indonesian New Order government's efforts to domesticate women.

4 *Abangan* refers to people who are nominally Muslim but who in fact embrace a more syncretic form of the religion which incorporates Hindu, animist and mystical (especially, Sufi) elements. It is characteristic of Java.

5 Non-governmental organisations, people's organisations, grassroots organisations, community-based organisations and voluntary organisations. Other acronyms include: DONGO (donor-organised non-governmental organisation); GONGO (government-organised non-governmental organisation); NGDO (non-governmental development organisation); NPO (non-profit organisation); PDO (private development organisation); PSO (public service organisation); PVO (private voluntary organisation); QANGO (quasi non-governmental organisation); and VALG (voluntary agency/organisation). Source: http://www.gdrc.org/ngo/ngo-ngdo-cbo.html.

6 NGOs are 'autonomous, non-membership, relatively permanent or institutionalised, non-profit (but not always voluntary) intermediary organisations, staffed by professionals or the educated elite, which work with grassroots organisations in a supportive capacity' (Desai 2002: 495).

7 See http://www.worldbank.org/participation/.

8 The intention here is not to provide a detailed account of these critical perspectives, but for more information see: Mohan and Stokke (2000), Cleaver (2001), Cooke and Kothari (2001), Mohan (2001) and Cornwall (2003).

9 McEwan (2005) also wonders whether participation in the South African context has been constructed as a political technology (see Table 7.3).

10 It should not be confused either with de-concentration, which involves the relocation of central government offices to regional or provincial levels while maintaining the original role of central government, nor with privatisation, in which responsibility and power is transferred from the public to the private sphere.

11 This is not the intention of the programme. Richer users are meant to subsidise poorer users. In practice, though, this can be difficult to achieve when authority is passed down to local groups.

12 This emerges from reading Tania Li's (1999, 2000, 2001, 2005) articles on Indonesia and Thomas' (1994) book on colonial Fiji (and elsewhere).

8 Alternatives: the everyday and resistance

1 The volumes, titled *Subaltern studies: writings on South Asian history and society*, were all published by Oxford University Press in New Delhi. Volumes 1–6 edited by Ranajit Guha (1982, 1983, 1984, 1985, 1987, 1989), volume 7 by Partha Chatterjee and Gyanendra Pandey (1992), volume 8 by David Arnold and David Hardiman ('Essays in honour of Ranajit Guha') (1994), volume 9 by Shahid Amin and Dipesh Chakrabarty (1996), and volume 10 by Gautam Bhadra, Gyan Prakash and Susie Tharu (1999).

2 For other territorialisation/counter-territorialisation studies see Peluso (1995), Vandergeest and Peluso (1995), Vandergeest (1996), Buch-Hansen (2003) and Wadley (2003).

3 It is weird on religious grounds because according to Buddhist norms trees cannot be ordained; only humans can be ordained. In 1996, in honour of the fiftieth anniversary of the King of Thailand's accession to the throne, the Northern Farmers' Network proposed ordaining 50 million trees – showing how, over a period of less than a decade, a marginal activity had become mainstream.

4 'Old' social movements, characteristic of Europe and North America – such as the trades union movement – were usually based on class and class interests. 'New' social movements such as the environmental, women's or peace movements represent, as Offe (1985: 833) says, the politics *of* a class but do not pursue politics *on behalf of* a class. In the South, however, new social movements are more likely to be (in broad terms) class-based and also, importantly, to be place-based. As is explored later in this chapter, new social movements in the South are associated with, for example, the interests of peasants and minority tribal groups.

5 Because appeals reflect poorly on local officials, local officials have in their turn tried to make it difficult or unattractive for people to appeal, by raising the risks and the costs of such action (Cai 2004).

6 For additional background on the study area see Butz (1995, 1996), and the discussion later in this chapter.

7 See, for example, Robins (2002), who explores the role of the eighteenth century English East India Company as the world's first transnational corporation.

8 An ADB commissioned environmental impact assessment of the project can be downloaded from http://www.adb.org/Documents/Environment/Ino/ino-central-sulawesi.pdf.

9 The structures of the everyday

1

> [W]hat I am . . . suggesting is that it might be possible to approach the production of place and culture not only from the side of the global, but of the local; not from the perspective of its abandonment but of its critical affirmation; not only according to the flight from places, whether voluntary or forced, but of the attachment to them.
>
> (Escobar 2001: 147–148)

References

Abdelali-Martini, Malika, Goldey, Patricia and Bailey, Elizabeth (2003) 'Towards a feminization of agricultural labour in Northwest Syria', *Journal of Peasant Studies* 30(2): 71–94.

Agarwal, Bina (2000) 'Conceptualising environmental collective action: why gender matters', *Cambridge Journal of Economics* 24: 283–310.

Agarwal, Bina (2001) 'Participatory exclusions, community forestry, and gender: an analysis for South Asia and a conceptual framework', *World Development* 29(10): 1623–1648.

Agnew, John A. and Duncan, James S. (1989) 'Introduction', in: John A. Agnew and James S. Duncan (eds) *The power of place: bringing together geographical and sociological imaginations*, Boston, MA: Unwin Hyman, pp. 1–8.

Alexander, Jennifer and Alexander, Paul (2004) 'Labour practices outside the factory: modern forms of household production in Java', in: Rebecca Elmhirst and Ratna Saptari (eds) *Labour in Southeast Asia: local processes in a globalised world*, London: Routledge, pp. 215–234.

Amin, Ash (2002) 'Spatialities of globalization', *Environment and Planning A* 34: 385–399.

Appadurai, Arjun (1999) 'Globalization and the research imagination', *International Social Science Journal* 51(160): 229–238.

Appadurai, Arjun (2000) 'Grassroots globalization and the research imagination', *Public Culture* 12(1): 1–19.

Arce, Alberto (2003) 'Value contestations in development interventions: community development and sustainable livelihoods approaches', *Community Development Journal* 38(3): 199–212.

Archer, Margaret S. (1995) *Realist social theory: the morphogenetic approach*, Cambridge: Cambridge University Press.

Armitage, John, Bishop, Ryan, and Kellner, Douglas (2005) 'Introducing cultural politics', *Cultural Politics* 1(1): 1–4.

Arnold, David (2005) 'Agriculture and "improvement" in early colonial India: a pre-history of development', *Journal of Agrarian Change* 5(4): 505–525.

Bangkok Post (1998) 'Back to relying on basics', *Bangkok Post Year End Review*, 15 January, pp. 8–9.

Bardhan, Pranab (2006) 'Globalization and rural poverty', *World Development* 34(8): 1393–1404.

Barkin, David (1990) *Distorted development: Mexico in the world economy*, Boulder, CO: Westview Press.

Barkin, David (2002) 'The reconstruction of a modern Mexican peasantry', *Journal of Peasant Studies* 30(1): 73–90.

Baulch, Bob and Hoddinott, John (2000) 'Economic mobility and poverty dynamics in developing countries', *Journal of Development Studies* 36(6): 1–24.

Beard, Victoria (1999) 'Navigating and creating spaces: an Indonesian community's struggle for land tenure', *Plurimondi* 1(2): 127–145.

Beard, Victoria (2002) 'Covert planning for social transformation in Indonesia', *Journal of Planning Education and Research* 22: 15–25.

Beard, Victoria (2003) 'Learning radical planning: the power of collective action', *Planning Theory* 2(1): 13–35.

Bebbington, Anthony (2002) 'Sharp knives and blunt instruments: social capital in development studies', *Antipode* 34(4): 800–803.

Bebbington, Anthony (2003) 'Global networks and local developments: agendas for development geography', *Tijdschrift voor Economische en Sociale Geografie* 94(3): 297–309.

Bebbington, Anthony (2004a) 'Movements and modernizations, markets and municipalities: indigenous federations in rural Ecuador', in: Richard Peet and Michael Watts (eds) *Liberation ecologies: environment, development, social movements*, London: Routledge, pp. 394–421.

Bebbington, Anthony (2004b) 'Social capital and development studies 1: critique, debate, progress?', *Progress in Development Studies* 4(4): 343–349.

Bebbington, Antony (2005) 'Donor–NGO relations and representations of livelihood in non-governmental aid chains', *World Development* 33(6): 937–950.

Bebbington, Anthony and Perreault, Thomas (1999) 'Social capital, development, and access to resources in highland Ecuador', *Economic Geography* 75(4): 395–418.

Bebbington, Antony, Dharmawan, Leni, Fahmi, Erwin, and Guggenheim, Scott (2004) 'Village politics, culture and community-driven development: insights from Indonesia', *Progress in Development Studies* 4(3): 187–205.

Becker, Gary S. (1991 [1981]) *A treatise on the family*, enlarged edition, Cambridge, MA: Harvard University Press.

Begum, Sharifa and Sen, Binayak (2004) *Unsustainable livelihoods, health shocks and urban chronic poverty: rickshaw pullers as a case study*, Chronic Poverty Research Centre Working Paper 46, Bangladesh Institute of Development Studies, Dhaka (November). Downloaded from http://www.chronicpoverty.org/pdfs/46%20Begum_Sen.pdf.

Begum, Sharifa and Sen, Binayak (2005) 'Pulling rickshaws in the city of Dhaka: a way out of poverty?', *Environment and Urbanization* 17(2): 11–25.

Berger, Mark T. (2004) 'After the Third World? History, destiny and the fate of Third Worldism', *Third World Quarterly* 25(1): 9–39.

Bhagwati, Jagdish (1958) 'Immiserizing growth: a geometrical note', *Review of Economic Studies* 25(3): 201–205.

Bhagwati, Jagdish (2004) *In defence of globalization*, New York: Oxford University Press.

Black, Richard and Watson, Elizabeth (2006) 'Local community, legitimacy and cultural authenticity in postconflict natural resource management: Ethiopia and Mozambique', *Environment and Planning D: Society and Space* 24: 263–282.

Bonnett, Alastair (2003) 'Geography as the world discipline: connecting popular and academic geographical imaginations', *Area* 35(1): 55–63.

Bourdieu, Pierre (1985) 'The forms of capital', in: John Richardson (ed.) *Handbook of theory and research for the sociology of education*, New York: Greenwood, pp. 241–258.

Bourdieu, Pierre (1986) *Distinction: a social critique of the judgment of taste*, London: Routledge and Kegan Paul.

Bourdieu, Pierre (2000) 'Making the economic habitus: Algerian workers revisited', *Ethnography* 1(1): 17–41.

Bowden, Bob (2002) 'Young people, education and development', in: Vandana Desai and Robert B. Potter (eds) *The companion to development studies*, London: Arnold, pp. 405–409.

Brandt, Willy (1980) *North–South: a programme for survival*, London: Pan.

Brass, Tom (2002a) 'On which side of what barricade? Subaltern resistance in Latin America and elsewhere', *Journal of Peasant Studies* 29(3): 336–399.

Brass, Tom (2002b) 'Latin American peasants – new paradigms for old?', *Journal of Peasant Studies* 29(3): 1–40.

Breman, J. (1995) 'Work and life of the rural proletariat in Java's coastal plain', *Modern Asian Studies* 29(1): 1–44.

Breman, J. and Wiradi, G. (2002). *Good times and bad times in rural Java: case study of socio-economic dynamics in two villages towards the end of the twentieth century*. Leiden: KITLV Press.

Brettell, Caroline B. (2002) 'The individual/agent and culture/structure in the history of the social sciences', *Social Science History* 26(3): 429–445.

Broad, Robin and Heckscher, Zahara (2003) 'Before Seattle: the historical roots of the current movement against corporate-led globalisation', *Third World Quarterly* 24: 713–728

Brocklesby, Mary Ann and Fisher, Eleanor (2003) 'Community development in sustainable livelihoods approaches – an introduction', *Community Development Journal* 38(3): 185–198.

Bruner, Edward M. (1961) 'Urbanization and ethnic identity in North Sumatra', *American Anthropologist* 63: 508–521.

Buch-Hansen, Mogens (2003) 'The territorialisation of rural Thailand: between localism, nationalism and globalism', *Tijdschrift voor Economische en Sociale Geografie* 94(3): 322–334.

Butz, David (1995) 'Legitimating porter regulation in an indigenous mountain community in northern Pakistan', *Environment and Planning D: Society and Space* 13(4): 381–414.

Butz, David (1996) 'Sustaining indigenous communities: symbolic and instrumental dimensions of pastoral resource use in Shimshal, northern Pakistan', *The Canadian Geographer* 40(1): 36–53.

Butz, David (2002) 'Resistance, representation and third space in Shimshal village, northern Pakistan', *Acme* 1: 15–34.

Cai, Yongshun (2004) 'Managed participation in China', *Political Science Quarterly* 119(3): 425–451.

Camfield, L., Choudhury, K. and Devine, J. (2006) *Relationships, happiness and well-being: insights from Bangladesh*, WeD Working Paper no. 14, ESRC Research Group on Wellbeing in Development Countries, Bath, UK. Downloaded from http://www.bath.ac.uk/econ-dev/wellbeing/research/workingpaperpdf/wed14.pdf.

Chakrabarty, Dipesh (2000) *Provincialising Europe: postcolonial thought and historical difference*, Princeton, NJ: Princeton University Press.

Chamberlain, James R., Alton, Charles, and Crisfield, Arthur G. (1995) *Indigenous peoples profile: Lao People's Democratic Republic*, CARE International, Vientiane (prepared for the World Bank, December).

Chambers, Robert (1988) *Sustainable livelihoods, environment and development: putting poor rural people first*, IDS Discussion Paper 240, Brighton, Sussex: Institute of Development Studies. Downloadable from http://www.ids.ac.uk/ids/bookshop/wp.html.

Chambers, Robert (1997) *Whose reality counts? Putting the first last*, London: Intermediate Technology.

Chambers, Robert (2004) *Ideas for development: reflecting forwards*, IDS Working Paper 238, Brighton, Sussex: Institute of Development Studies.

Chambers, Robert and Conway, Gordon (1992) *Sustainable rural livelihoods: practical concepts for the 21st century*, IDS Discussion Paper 296, Brighton, Sussex: Institute of Development Studies. Downloadable from http://www.ids.ac.uk/ids/bookshop/dp/dp296.pdf.

Chant, Sylvia (1996) 'Women's roles in recession and economic restructuring in Mexico and the Philippines', *Geoforum* 27(3): 297–327.

Chant, Sylvia (1998) 'Household, gender and rural–urban migration: reflections on linkages and considerations for policy', *Environment and Urbanization* 10(1): 5–21.

Chapman, Murray (1995) 'Island autobiographies of movement: alternative ways of knowing?', in: P. Claval and Singaravelou (eds) *Ethnogéographies*, Paris: L'Harmattan, pp. 247–259.

Chatthip Nartsupha (1999 [1984]) *The Thai village economy in the past*, translation by Chris Baker and Pasuk Phongpaichit, Chiang Mai, Thailand: Silkworm.

Chhotray, Vasudha (2004) 'The negation of politics in participatory development projects, Kurnool, Andhra Pradesh', *Development and Change* 35(2): 327–352.

Cleary, Mark C. (1996) 'Indigenous trade and European economic intervention in North-West Borneo c.1860–1930', *Modern Asian Studies* 30(2): 301–324.

Cleary, Mark C. (1997) 'From hornbills to oil? Patterns of indigenous and European trade in colonial Borneo', *Journal of Historical Geography* 23(1): 29–45.

Cleaver, Frances (2001) 'Institutions, agency and the limitations of participatory approaches to development', in: Bill Cooke and Uma Kothari (eds) *Participation: the new tyranny?*, London: Zed Books, pp. 36–55.

Cliggett, Lisa (2003) 'Gift remitting and alliance building in Zambian modernity: old answers to modern problems', *American Anthropologist* 105(3): 543–552.

Cohen, Margot (1996) 'Twisting arms for alms', *Far Eastern Economic Review*, 2 May, pp. 25–29.

Coleman, James S. (1988) 'Social capital and the creation of human capital', *American Journal of Sociology* 94(Supplement): S95–S120.

Coleman, James S. (1990) *Foundatons of social theory*, Cambridge, MA: Harvard University Press.

Conticini, Alessandro (2005) 'Urban livelihoods from children's perspectives: protecting and promoting assets on the streets of Dhaka', *Environment and Urbanization* 17(2): 69–81.

Cooke, Bill and Kothari, Uma (eds) (2001) *Participation: the new tyranny?*, London: Zed Books.

Corbridge, Stuart and Jewitt, Sarah (1997) 'From forest struggles to forest citizens? Joint Forest Management in the unquiet woods of India's Jharkhand', *Environment and Planning A* 29(12): 2145–2164.

Cornwall, Andrea (2002) 'Locating citizenship participation', *IDS Bulletin* 33(2): 49–58.

Cornwall, Andrea (2003) 'Whose voices? Whose choices? Reflections on gender and participatory development', *World Development* 31(8): 1325–1342.

CPD (2003) *Child labour policy in Bangladesh: what are we looking for?*, Report no. 61, Centre for Policy Dialogue, Dhaka, Bangladesh (July). Downloadable from http://www.cpd-bangladesh.org/publications/dr/DR-61.pdf.

Crehan, Kate (1997) *The fractured community: landscapes of power and gender in rural Zambia*, Berkeley, CA: University of California Press.

Curry, George and Koczberski, Gina (1998) 'Migration and circulation as a way of life for the Wosera Abelam of Papua New Guinea', *Asia Pacific Viewpoint* 39(1): 29–52.

Curry, George and Koczberski, Gina (1999) 'The risks and uncertainties of migration: an exploration of recent trends amongst the Wosera Abelam of Papua New Guinea', *Oceania* 70: 130–145.

Curtis, Fred (2004) 'Eco-localism and sustainability', *Ecological Economics* 46: 83–102.

Dannecker, Petra (2005) 'Bangladeshi migrant workers in Malaysia: the construction of the "Others" in a multi-ethnic context', *Asian Journal of Social Science* 33(2): 246–267.

Darunee Jongudomkarn and Camfield, Laura (2005) *Exploring the quality of life of people in North Eastern and Southern Thailand*, Working Paper of the ESRC Research Group on Wellbeing in Developing Countries, University of Bath, Bath, UK. Downloaded from www.welldev.org.uk.

Dasgupta, Biplab (1978) 'Introduction', in: Biplab Dasgupta (ed.) *Village studies in the Third World*, Delhi: Hindustan Publishing Corporation, pp. 1–12.

de Bruijn, Mirjam and van Dijk, Han (2003) 'Changing population mobility in West Africa: Fulbe pastoralists in central and south Mali', *African Affairs* 102: 285–307.

de Haan, Arjan (1999) 'Livelihoods and poverty: the role of migration – a critical review of the migration literature', *Journal of Development Studies* 36(2): 1–47.

de Haan, Arjan (2002) 'Migration and livelihoods in historical perspective: a case study of Bihar, India', *Journal of Development Studies* 38(5): 115–142.

de Haan, Arjan (2004) 'Calcutta's labour migrants: encounters with modernity', in: Katy Gardner and Filippo Osella (eds) *Migration, modernity and social transformation in South Asia*, Contributions to Indian Sociology, Occasional Studies 11, New Delhi: Sage, pp. 189–215.

de Haan, Arjan and Rogaly, Ben (2002) 'Introduction: migrant workers and their role in rural change', *Journal of Development Studies* 38(5): 1–14.

de Haan, Leo and Zoomers, Annelies (2003) 'Development geography at the crossroads of livelihood and globalisation', *Tijdschrift voor Economische en Sociale Geografie* 94(3): 350–362.

de Haan, Leo and Zoomers, Annelies (2005) 'Exploring the frontiers of livelihood research', *Development and Change* 36(1): 27–47.

de Neve, Geert (2004) 'Expections and rewards of modernity: commitment and mobility among rural migrants in Tirupur, Tamil Nadu', in: Katy Gardner and Filippo Osella (eds) *Migration, modernity and social transformation in South Asia*, Contributions to Indian Sociology, Occasional Studies 11, New Delhi: Sage, pp. 252–280.

Dercon, Stefan and Krishnan, Pramila (2000) *Poverty and survival strategies in Ethiopia during economic reform*, Research Report ESCOR no. 7280 (December), London: Department for International Development.

Desai, Vandana (2002) 'Role of non-governmental organizations (NGOs)', in: Vandana Desai and Robert B. Potter (eds) *The companion to development studies*, London: Arnold, pp. 495–499.

Deshingkar, Priya and Start, Daniel (2003) *Seasonal migration for livelihoods in India: coping, accumulation and exclusion*, Working Paper 220, London: Overseas Development Institute. Downloaded from http://www.odi.org.uk/publications/working_papers/wp220.pdf.

DFID (1999) *Social capital*, Key sheets for sustainable livelihoods 3, London: Department for International Development. Downloaded from http://www.keysheets.org/red_3_Soccap_rev.pdf.

DFID (2004) *DFID rural and urban development case study – Bangladesh*, prepared for the Department for International Development by Oxford Policy Management (July). Downloadable from http://passlivelihoods.org.uk/site_files/files/reports/project_id_167/Bangladesh%20Rural%20Urban%20Change%20Case%20Study_RU0173.pdf.

Dicken, Peter (2004) 'Geographers and "globalization": (yet) another missed boat?', *Transactions of the Institute of British Geographers NS* 29: 5–26.

Dicken, Peter (2007 [1986]) *Global shift: mapping the changing contours of the world economy*, 5th edition, London: Sage.

Duncan, James S. (2002) 'Embodying colonialism? Domination and resistance in nineteenth-century Ceylonese coffee planatations', *Journal of Historical Geography* 28(3): 317–338.

Economist, The (2006a) 'The new powers in giving', *The Economist*, 1 July: 69–71.

Economist, The (2006b) 'The new titans: a survey of the world economy', *The Economist*, 16 September.

Economist, The (2006c) 'Big questions, big numbers', special report on economic models, *The Economist*, 15 July: 75–77.

Edgerton, Robert B. (1985) *Rules, exceptions and social order*, Berkeley, CA: University of California Press.

Edwards, Michael (2000) 'More social capital, less global poverty?', *Development Outreach*, World Bank Institute (Summer). Downloaded from http://www1.worldbank.org/devoutreach/summer00/article.asp?id=67.

Elden, Stuart (2004) 'Between Marx and Heidegger: politics, philosophy and Lefebvre's *The production of space*', *Antipode* 36(1): 86–105.

Eliot, Joshua, Capaldi, Liz, and Bickersteth, Jane (2001) *Indonesia handbook*, Bath: Footprint.

Ellis, Frank (1999) 'Rural livelihood diversity in developing countries: evidence and policy implications', *Natural Resource Perspectives no. 40*, London: Overseas Development Institute.

Elmhirst, Rebecca (1995a) 'Gender, environment and transmigration: comparing migrant and *pribumi* household strategies in Lampung, Indonesia', paper presented to the Third WIVS conference on *Indonesian Women in the Household and Beyond*, Royal Institute of Linguistics and Anthropology, Leiden, 25–29 September.

Elmhirst, Rebecca (1995b) '*Anak mas, anak tiri*: difference and convergence in the livelihood strategies of transmigrants and indigenous people in Indonesia', an outline of provisional findings presented to ICRAF, Bogor, Indonesia, 18 April.

Elmhirst, Rebecca (1996) 'Transmigration and local communities in North Lampung: exploring identity politics and resource control in Indonesia', paper presented at the Association of South East Asian Studies' (ASEASUK) Conference, School of Oriental and African Studies, London, 25–27 April.

Elmhirst, Rebecca (1997) 'Gender, environment and culture: a political ecology of transmigration in Indonesia', unpublished PhD thesis, University of London.

Elmhirst, Rebecca (1998a) '"*Kismon*" and "*kemarau*": a downward sustainability spiral in North Lampung tanslok settlement', report to the International Centre for Agroforestry Research (ICRAF), Bogor, Indonesia, May.

Elmhirst, Rebecca (1998b) 'Daughters and displacement: migration dynamics in an Indonesian transmigration area', paper presented at the workshop on migration and sustainable livelihoods, University of Sussex, June.

Elmhirst, Rebecca (1998c) 'Gender, culture and space: a political geography of factory labour in Indonesia', paper presented at the European South East Asian Studies (EUROSEAS) conference, Hamburg, 3–6 September.

Elmhirst, Rebecca (1999) 'Space, identity politics and resource control in Indonesia's transmigration programme', *Political Geography* 18: 813–835.

Elmhirst, Rebecca (2000) 'Negotiating gender, kinship and livelihood practices in an Indonesian transmigrant area', in Juliette Koning, Marleen Nolten, Janet Rodenburg and Ratna Saptari (eds) *Women and households in Indonesia: cultural notions and social practices*, Richmond, Surrey: Curzon, pp. 208–234.

Elmhirst, Rebecca (2001) 'Resource struggles and the politics of place in North Lampung, Indonesia', *Singapore Journal of Tropical Geography* 22(3): 284–306.

Elmhirst, Rebecca (2002) 'Daughters and displacement: migration dynamics in an Indonesian transmigration area', *Journal of Development Studies* 38(5): 138–166.

Elmhirst, Rebecca (2004) 'Labour politics in migrant communities: ethnicity and women's activism in Tangerang, Indonesia', in: Rebecca Elmhirst and Ratna Saptari (eds) *Labour in Southeast Asia: local processes in a globalised world*, London: Routledge, pp. 387–406.

Englund, Harri and Leach, James (2000) 'Ethnography and the meta-narratives of modernity', *Current Anthropology* 41(2): 225–248.

Environment and Urbanization (1998) Themed issue 'Beyond the rural–urban divide', *Environment and Urbanization* 10(1).

Environment and Urbanization (2003) Themed issue 'Rural–urban transformations', *Environment and Urbanization* 15(1).

Esara, Pilapa (2004) '"Women will keep the household": the mediation of work and family by female labor migrants in Bangkok', *Critical Asian Studies* 36(2): 199–216.

Escobar, Arturo (1995) *Encountering development: the making and unmaking of the Third World*, Princeton, NJ: Princeton University Press.

Escobar, Arturo (2001) 'Culture sits in places: reflections on globalism and subaltern strategies of localization', *Political Geography* 20(2): 139–174.

Escobar, Arturo (2004) 'Beyond the Third World: imperial globality, global coloniality and anti-globalisation social movements', *Third World Quarterly* 25(1): 207–230.

Evans, Hugh Emrys (1992) 'A virtuous circle model of rural-urban development: evidence from a Kenyan small town and its hinterland', *Journal of Development Studies* 28(4): 640–667.

Evans, Hugh Emrys and Ngau, Peter (1991) 'Rural–urban relations, household income diversification, and agricultural productivity', *Development and Change* 22: 519–545.

Fan, C. Cindy (2001) 'Migration and labor-market returns in urban China: results from a recent survey in Guangzhou', *Environment and Planning A* 33: 479–508.

Fan, C. Cindy (2002) 'The elite, the natives, and the outsiders: migration and labor market segmentation in urban China', *Annals of the Association of American Geographers* 92(1): 103–124.

Fan, C. Cindy (2003) 'Rural–urban migration and gender division of labor in transitional China', *International Journal of Urban and Regional Research* 27(1): 24–47.

Farrington, John, Carney, Diana, Ashley, Caroline, and Turton, Cathryn (1999) 'Sustainable livelihoods in practice: early applications of concepts in rural areas', *Natural Resource Perspectives 42*, London: Overseas Development Institute (June). Downloaded from http://www.odi.org.uk/nrp/42.html.

Farrington, John, Christoplos, Ian and Kidd, Andrew D. with Beckman, Marlin (2002) *Extension, poverty and vulnerability: the scope for policy reform*, Working Paper 155, London: Overseas Development Institute. Downloaded from http://www.odi.org.uk/publications/wp155_a.pdf.

Ferguson, James (1990a) 'Mobile workers, modernist narratives: a critique of the historiography of transition on the Zambian Copperbelt [part one]', *Journal of Southern African Studies* 16(3): 385–412.

Ferguson, James (1990b) 'Mobile workers, modernist narratives: a critique of the historiography of transition on the Zambian Copperbelt [part two]', *Journal of Southern African Studies* 16(4): 603–621.

Ferguson, James (1994) 'Modernist narratives, conventional wisdoms, and colonial liberalism: reply to a straw man', *Journal of Southern African Studies* 20(4): 633–640.

Ferguson, James (1999) *Expectations of modernity: myths and meanings of urban life on the Zambian copperbelt*, Berkeley, CA: University of California Press.

Fine, Ben (2002) 'They f**k you up those social capitalists', *Antipode* 34(4): 796–799.

Flint, C. (2002) 'Political geography: globalization, metapolitical geographies and everday life', *Progress in Human Geography* 26(3): 391–400.

Folbre, Nancy (1986a) 'Cleaning house: new perspectives on households and economic development', *Journal of Development Economics* 22(1): 5–40.

Folbre, Nancy (1986b) 'Hearts and spades: paradigms of household economics', *World Development* 14(2): 245–255.

Foucault, M. (1990) *The history of sexuality, Volume I: An Introduction*, translated by Robert Hurley, New York: Random House.

Francis, Paul (2001) 'Participatory development at the World Bank: the primacy of process', in: Bill Cooke and Uma Kothari (eds) *Participation: the new tyranny?*, London: Zed Books, pp. 72–87.

Francis, Paul (2002) *Social capital at the World Bank: strategic and operational implications of the concept*, Social Development Strategy Paper, Washington DC: World Bank. Downloadable from http://lnweb18.worldbank.org/ESSD/essdext.nsf/62DocByUnid/859A26E9E5E3400B85256C5500534E34/$FILE/Francis4.pdf.

Freidberg, Susanne (2003) 'French beans for the masses: a modern historical geography of food in Burkina Faso', *Journal of Historical Geography* 29(3): 445–463.

Friedman, Thomas (2005) *The world is flat: a brief history of the globalized world in the twenty-first century*, London: Allen Lane.

Friedmann, Harriet (1993) 'The political economy of food: a global crisis', *New Left Review* 197 (Jan.–Feb.): 29–57.

Gaiha, Raghav and Deolalikar, Anil B. (1993) 'Persistent, expected and innate poverty: estimates from semi-arid rural South India, 1975–1984', *Cambridge Journal of Economics* 17(4): 409–421.

García-Guadilla, María Pilar (2002) 'Democracy, decentralization, and clientelism: new relationships and old practices', *Latin American Perspectives* 29(5): 90–109.

Gardner, Katy and Osella, Filippo (2004) 'Migration, modernity and social transformation in South Asia: an introduction', in: Katy Gardner and Filippo Osella (eds) *Migration, modernity and social transformation in South Asia*, Contributions to Indian Sociology, Occasional Studies 11, New Delhi: Sage, pp. xi–xlviii.

Garrett, James and Chowdhury, Shyamal (2004) *Urban–rural links and transformation in Bangladesh: a review of the issues*, International Food Policy Research Institute (IFPRI) , Washington, DC (July), discussion paper prepared for Rural Livelihood Program (RLP), Care, Bangladesh. Downloaded from http://www.livelihoods.org/hot_topics/docs/UR_CARE.pdf.

Gasper, Des (2004) *Subjective and objective well-being in relation to economic inputs: puzzles and responses*, WeD (Wellbeing in Developing Countries) Working Paper 09 (October).

Giddens, Anthony (1984) *The constitution of society: outline of the theory of structuration*, Cambridge: Polity Press.

Giddens, Anthony (1990) *The consequences of modernity*, Stanford, CA: Stanford University Press.

Gill, Gerard (2003) *Seasonal labour migration in rural Nepal: a preliminary overview*, Working Paper 218, London: Overseas Development Institute. Downloaded from http://www.odi.org.uk/publications/working_papers/WP218.pdf.

Ginsburg, Norton (1991) 'Extended metropolitan regions in Asia: a new spatial paradigm', in: Norton Ginsburg, Bruce Koppel and T.G. McGee (eds) *The extended metropolis: settlement transition in Asia*, Honolulu: University of Hawaii Press, pp. 27–46.

Glassman, Jim (2002) 'From Seattle and Ubon to Bangkok: the scales of resistance to corporate globalization', *Environment and Planning D: Society and Space* 20: 513–533.

Gough, Katherine and Yankson, Paul W.K. (2000) 'Land markets in African cities: the case of peri-urban Accra, Ghana', *Urban Studies* 37(13): 2485–2500.

Gramsci, A. (1971) *Selections from the prison notebooks*, translated and edited by Q. Hoare and G.N. Smith, London: Lawrence and Wishart.

Grant, Richard and Nijman, Jan (2002) 'Globalization and the corporate geography of cities in the less-developed world', *Annals of the Association of American Geographers* 92(2): 320–340.

Grijns, Mies and van Velzen, Anita (1993) 'Working women: differentiation and marginalisation', in: Chris Manning and Joan Hardjono (eds) *Indonesia assessment 1993 – Labour: sharing in the benefits of growth?*, Political and Social Change Monograph no. 20, Research School of Pacific Studies, Canberra: Australian National University, pp. 214–228.

Guha, Ranajit (1982a) 'On some aspects of the historiography of colonial India', in: Ranajit Guha (ed.) *Subaltern studies I: writings on South Asian history and society*, New Delhi: Oxford University Press, pp. 1–8.

Guha, Ranajit (1982b) 'Preface', in: Ranajit Guha (ed.) *Subaltern studies I: writings on South Asian history and society*, Delhi: Oxford University Press, pp. vii–viii.

Guha, Ranajit (1998) 'Introduction', in: Ranajit Guha (ed.) *A subaltern studies reader, 1986–1995*, Delhi: Oxford University Press, pp. ix–xxii

Guinness, Patrick (1994) 'Local society and culture', in: Hal Hill (ed.) *Indonesia's New Order: the dynamics of socio-economic transformation*, Honolulu: University of Hawaii Press, pp. 267–304.

Gutiérrez, Javier and López-Nieva, Pedro (2001) 'Are international journals of human geography really international?', *Progress in Human Geography* 25(1): 53–69.

Guyer, Jane I. (1981) 'Household and community in African studies', *African Studies Review* 24(2–3): 87–137.

Hale, Angela and Opondo, Maggie (2005) 'Humanising the cut flower chain: confronting the realities of flower production for workers in Kenya', *Antipode* 37(2): 301–322.

Hamilton, Sarah and Fischer, Edward (2003) 'Non-traditional agricultural exports in highland Guatemala: understandings of risk and perceptions of change', *Latin American Research Review* 38(3): 82–110.

Hampshire, Kate (2002) 'Fulani on the move: seasonal economic migration in the Sahel as a social process', *Journal of Development Studies* 38(5): 15–36.

Hancock, Peter (2000) 'Women workers still exploited: revisiting two Nike factories in West Java after the economic crisis', *Inside Indonesia* 62 (April–June): 21–22.

Hancock, Peter (2001) 'Rural women earning income in Indonesian factories: the impact on gender relations', *Gender and Development* 9(1): 18–24.

Harriss, John and Renzio, Paolo de (1997) 'Missing link or analytically missing? The concept of social capital', *Journal of International Development* 9(7): 919–937.

Hart, Gillian (1991) 'Engendering everyday resistance: gender, patronage and production politics in rural Malaysia', *Journal of Peasant Studies* 19(1): 93–121.

Hasan, Arif (2002) 'The changing nature of the informal sector in Karachi as a result of global restructuring and liberalization', *Environment and Urbanization* 14(1): 69–78.

Hayami, Y. and Kikuchi, M. (1981) *Asian village economy at the crossroads: an economic approach to institutional change*, Tokyo: University of Tokyo Press.

Hettige, Hemamala (2006) *When do rural roads benefit the poor and how? An in-depth analysis based on case studies*, Operations Evaluation Department, Manila: Asian Development Bank. Downloaded from http://www.adb.org/Documents/Books/ruralroad_benefits/rural-roads.pdf.

Hewison, Kevin (2001) 'Nationalism, populism, dependency: Southeast Asia and responses to the Asian crisis', *Singapore Journal of Tropical Geography* 22(3): 219–236.

Hossain, Mahabub (2004) *Poverty alleviation through agriculture and rural development in Bangladesh*, Paper no. 39, Centre for Policy Dialogue, Dhaka, Bangladesh (July). Downloadable from http://www.cpd-bangladesh.org/publications/op/OP39.pdf.

Huntington, Ellsworth (1924) *Climate and civilization*, 3rd edition, New Haven, CT: Yale University Press.

Ingham, Barbara (1993) 'The meaning of development: interactions between "new" and "old" ideas', *World Development* 21(11): 1803–1821.

Inkeles, Alex and Smith, David H. (1974) *Becoming modern: individual change in six developing countries*, London: Heinemann.

Inkeles, Alex, Broaded, C. Montgomery and Zhongde Cao (1997) 'Causes and consequences of individual modernity in China', *The China Journal* 37: 31–59.

Isager, Lotte and Ivarsson, Søren (2002) 'Contesting landscapes in Thailand: tree ordination as counter-territorialization', *Critical Asian Studies* 34(3): 395–417.

Ives, J. and Messerli, B. (1989) *The Himalayan dilemma: reconciling development and conservation*, London: Routledge.

Iyenda, Guillaume (2005) 'Street enterprises, urban livelihoods and poverty in Kinshasa', *Environment and Urbanization* 17(2): 55–67.

Jackson, Jean E. (1995) 'Culture, genuine and spurious: the politics of Indianness in the Vaupés , Colombia', *American Ethnologist* 22(1): 3–27.

Jalan, Jyotsa and Ravallion, Martin (2000) 'Is transient poverty different? Evidence from China', *Journal of Development Studies* 36(6): 82–99.

James, Jeffrey (2006) 'Misguided investments in meeting Millennium Development Goals: a reconsideration using ends-based targets', *Third World Quarterly* 27(3): 443–458.

Johnson, Craig, Deshingkar, Priya and Start, Daniel (2005) 'Grounding the state: devolution and development in India's *panchayats*', *Journal of Development Studies* 41(6): 937–970.

Johnston, Ron (1985) 'The world is our oyster', *Transactions of the Institute of British Geographers* NS 9: 443–459.

Jonas, Andrew E.G. (2006) 'Pro scale: further reflections on the "scale debate" in human geography', *Transactions of the Institute of British Geographers* 31: 399–406.

Kabeer, Naila (1997) 'Women, wage and intra-household power relations in urban Bangladesh', *Development and Change* 28: 261–302.

Kahn, Joel S. (2001) 'Anthropology and modernity', *Current Anthropology* 42(5): 651–680.

Kaplinsky, Raphael (2000) 'Globalisation and unequalisation: what can be learned from value chain analysis?', *Journal of Development Studies* 37(2): 117–146.

Katz, Elizabeth (1995) 'Gender and trade within the household: observations from rural Guatemala', *World Development* 23(2): 327–342.

Kearney, M. (1996) *Reconceptualising the peasantry: anthropology in global perspective*, Boulder, CO: Westview Press.

Kelly, Philip (1999a) 'Everyday urbanization: the social dynamics of development in Manila's extended metropolitan region', *International Journal of Urban and Regional Research* 23(2): 283–303.

Kelly, Philip (1999b) 'Rethinking the "local" in labour markets: the consequences of cultural embeddedness in a Philippine growth zone', *Singapore Journal of Tropical Geography* 20(1): 56–75.

Kelly, Philip (1999c) 'Globalization, power and the politics of scale in the Philippines', *Geoforum* 28(2): 151–171.

Kelly, Philip (2000) *Landscapes of globalisation: human geographies of economic change in the Philippines*, London: Routledge.

Kerkvliet, Benedict J. (1995a) 'Rural society and state relations', in: Benedict J. Kerkvliet and Doug J. Porter (eds) *Vietnam's rural transformation*, Boulder, CO: Westview Press and Singapore: Institute of Southeast Asian Studies, pp. 65–96.

Kerkvliet, Benedict J. (1995b) 'Village–state relations in Vietnam: the effects of everyday politics on decollectivization', *Journal of Asian Studies* 54(2): 396–418.

Kerkvliet, Benedict J. (2005) *The power of everyday politics: how Vietnamese peasants transformed national policy*, Singapore: Institute of Southeast Asian Studies.

Koczberski, Gina and Curry, George (2004) 'Divided communities and contested landscapes: mobility, development and shifting identities in migrant destination sites in Papua New Guinea', *Asia Pacific Viewpoint* 45(3): 357–371.

Koizumi, Junko (1992) 'The commutation of Suai from Northeast Siam in the middle of the nineteenth century', *Journal of Southeast Asian Studies* 23(2): 276–307.

Kong, Lily (1999) 'Globalisation and Singaporean transmigration: re-imaging and negotiating national identity', *Political Geography* 18: 563–589.

Koning, Juliette (2005) 'The impossible return? The post-migration narratives of young women in rural Java', *Asian Journal of Social Science* 33(2): 165–185.

Korf, Benedikt (2004) 'War, livelihoods and vulnerability in Sri Lanka', *Development and Change* 35(2): 275–295.

Korf, Benedikt and Fünfgeld, Hartmut (2006) 'War and the commons: assessing the changing politics of violence, access and entitlements in Sri Lanka', *Geoforum* 37: 391–403.

Korf, Benedikt and Oughton, Elizabeth (2006) 'Rethinking the European countryside – can we learn from the South?', *Journal of Rural Studies* 22: 278–289.

Krishna, Anirudh, Kapila, Mahesh, Porwal, Mahendra, and Singh, Virpal (2005) 'Why growth is not enough: household poverty dynamics in Northeast Gujarat, India', *Journal of Development Studies* 41(7): 1163–1192.

Kritzinger, Andrienetta, Barrientos, Stephanie and Rossouw, Hester (2004) 'Global production and flexible employment in South African horticulture: experiences of contract workers in fruit exports', *Sociologia Ruralis* 44(1): 17–39.

Krugman, Paul (1997) 'In praise of cheap labor: bad jobs at bad wages are better than no jobs at all', *The dismal science*. Downloadable from http://web.mit.edu/krugman/www/smokey.html.

Kuhn, Randall (2004) 'Identities in motion: social exchange networks and rural-urban migration in Bangladesh', in: Katy Gardner and Filippo Osella (eds) *Migration, modernity and social transformation in South Asia*, Contributions to Indian Sociology, Occasional Studies 11, New Delhi: Sage, pp. 311–337.

Kunfaa, Ernest Y. with Lambongang, Joe, Dogbe, Tony, and MacKay, Heather (1999) *Consultations with the poor: Ghana country synthesis report*, Centre for the Development of People (CEDEP), Kumasi, Ghana (July). Downloadable from: http://www1.worldbank.org/prem/poverty/voices/reports.htm.

Kunfaa, Ernest Y. and Dogbe, Tony with MacKay, Heather J. and Marshall, Celia (2000) 'Ghana: "empty pockets"', in: Deepa Narayan and Patti Petesch (eds) *Voices of the poor: from many lands*, New York: Oxford University Press, pp. 17–49. Downloadable from http://www1.worldbank.org/prem/poverty/voices/reports.htm.

Lansing, Stephen (1991) *Priests and programmers: technologies of power in the engineered landscape of Bali*, Princeton, NJ: Princeton University Press.

Leach, Melissa, Mearns, Robin and Scoones, Ian (1999) 'Environmental entitlements: dynamics and institutions in community-based natural resource management', *World Development* 27(2): 225–247.

Lefebvre, Henri (1991) *Critique of everyday life*, London: Verso.

Leinbach, Thomas R. and Del Casino, Vincent J. Jr. (1998) 'The family mode of production and its fungibility in Indonesian transmigration: the example of Makarti Jaya, South Sumatra', *Sojourn* 13(2): 193–219.

Leinbach, Thomas R. and Watkins, John E. (1998) 'Remittances and circulation behaviour in the livelihood process: transmigrant families in South Sumatra, Indonesia', *Economic Geography* 74(1): 45–63.

Lemoine, Jacques (2002) *Wealth and poverty: a case study of the Kim Di Mun (Lantène Yao, Lao Houay) of the Nam Ma Valley, Meuang Long District, Louang Namtha, Lao PDR*, Working Papers on Poverty Reduction no. 10, Committee for Planning and Cooperation, National Statistics Center, Vientiane (December).

Ley, David (2004) 'Transnational spaces and everyday lives', *Transactions of the Institute of British Geographers NS* 29: 151–164.

Li, Tania (1999) 'Compromising power: development, culture and rule in Indonesia', *Cultural Anthropology* 14(3): 295–322.

Li, Tania (2000) 'Articulating indigenous identity in Indonesia: resource politics and the tribal slot', *Comparative Studies in Society and History* 42(1):149–179.

Li, Tania (2001) 'Masyarakat adat, difference, and the limits of recognition in Indonesia's forest zone', *Modern Asian Studies* 35(3): 645–676.

Li, Tania (2005) 'Beyond "the state" and failed schemes', *American Anthropologist* 107(3): 383–394.

Little, Peter D. and Watts, Michael J. (eds) (1994) *Living under contract: contract farming and agrarian transformation in sub-Saharan Africa*, Madison, WI: University of Wisconsin Press.

Long, Norman (2001) *Development sociology: actor perspectives*, London: Routledge.

Longwe, Sara Hlupekile (2000) 'Towards realistic strategies for women's political empowerment in Africa', *Gender and Development* 8(3): 24–30.

Lynch, Kenneth (2005) *Rural–urban interactions in the developing world*, London: Routledge.

Lyon, Fergus (2000) 'Trust, networks and norms: the creation of social capital in agricultural economies in Ghana', *World Development* 28(4): 663–682.

Lyon, Fergus (2003) 'Group co-operation and agricultural market access in Ghana', *Community Development Journal* 38(4): 323–331.

Lyons, Michal and Snoxell, Simon (2005) 'Sustainable urban livelihoods and marketplace social capital: crisis and strategy in petty trade', *Urban Studies* 42(8): 1301–1320.

Ma, Lawrence J.C. and Fan, Ming (1994) 'Urbanization from below: the growth of towns in Jiangsu, China', *Urban Studies* 31(10): 1625–1645.

McEwan, Cheryl (2003) '"Bringing government to the people": women, local governance and community participation in South Africa', *Geoforum* 34: 469–481.

McEwan, Cheryl (2005) 'New spaces of citizenship? Rethinking gendered participation and empowerment in South Africa', *Political Geography* 24: 969–991.

McGee, T.G. (1989) 'Urbanisasi or kotadesasi? Evolving patterns of urbanization in Asia', in: Frank J. Costa, Ashok K. Dutt, Lawrence J.C. Ma and Allen G. Noble (eds) *Urbanization in Asia: spatial dimensions and policy issues*, Honolulu: University of Hawaii Press, pp. 93–108.

McGee, T.G. (1991a) 'Eurocentrism in geography: the case of Asian urbanization', *The Canadian Geographer* 35(4): 332–344.

McGee, T.G. (1991b) 'The emergence of desakota regions in Asia: expanding a hypothesis', in: Norton Ginsburg, Bruce Koppel and T.G. McGee (eds) *The extended metropolis: settlement transition in Asia*, Honolulu: University of Hawaii Press, pp. 3–25.

McGee, T.G. and Greenberg, Charles (1992) 'The emergence of extended metropolitan regions in ASEAN', *ASEAN Economic Bulletin* 9(1): 22–44.

McGregor, J. Allister (2007) 'Researching wellbeing: from concepts to methodology', in: Ian Gough and J. Allister McGregor (eds) *Wellbeing in developing countries: from theory to research*, Cambridge: Cambridge University Press.

McKay, Deidre (2003) 'Cultivating new local futures: remittance economies and land-use patterns in Ifugao, Philippines', *Journal of Southeast Asian Studies* 34(2): 285–306.

McKay, Deidre (2005) 'Reading remittance landscapes: female migration and agricultural transition in the Philippines', *Geografisk Tidsskrift, Danish Journal of Geography* 105(1): 89–99.

McKay, Deidre and Brady, Carol (2005) 'Practices of place-making: globalisation and locality in the Philippines', *Asia Pacific Viewpoint* 46(2): 89–103.

Macmillan, Hugh (1993) 'The historiography of transition on the Zambian Copperbelt – another view', *Journal of Southern African Studies* 19(4): 681–712.

Macmillan, Hugh (1996) 'More thoughts on the historiography of transition on the Zambian Copperbelt', *Journal of Southern African Studies* 22(2): 309–312.

McSweeney, Kendra (2004) 'The dugout canoe trade in Central America's Mosquitia: approaching rural livelihoods through systems of exchange', *Annals of the Association of American Geographers* 94(3): 638–661.

Mallon, Florencia E. (1994) 'The promise and dilemma of subaltern studies: perspectives from Latin America', *American Historical Review* 99(5): 1491–1515.

Mansfield, Becky (2005) 'Beyond rescaling: reintegrating the "national" as a dimension of scalar relations', *Progress in Human Geography* 29(4): 458–473.

Marston, Sallie A., Jones III, John Paul, and Woodward, Keith (2005) 'Human geography without scale', *Transactions of the Institute of British Geographers* 30: 416–432.

Martin, Ron (2004) 'Geography: making a difference in a globalizing world', *Transactions of the Institute of British Geographers NS* 29: 147–50.

Mawdsley, Emma, Townsend, Janet, Porter, Gina, and Oakley, Peter (2002) *Knowledge, power and development agendas: NGOs North and South*, Oxford: INTRAC.

May, Lord (2005) 'Threats to tomorrow's world', Anniversary Address to the Royal Society, London. Downloaded from http://www.royalsoc.ac.uk/downloaddoc.asp?id=2414.

Mazlish, Bruce (1991) 'The breakdown of connections and modern development', *World Development* 19(1): 31–44.

Mearns, Robin (2004) 'Sustaining livelihoods on Mongolia's pastoral commons: insights from a participatory poverty assessment', *Development and Change* 35(1): 107–139.

Merrifield, Andrew (1993) 'Place and space: a Lefebvrian reconciliation', *Transactions of the Institute of British Geographers* New Series, 18(4): 516–531.

Merrifield, Andrew (2000) 'Henri Lefebvre: a socialist in space', in: Mike Crang and Nigel Thrift (eds) *Thinking space*, London: Routledge, pp. 167–182.

Michener, Victoria J. (1998) 'The participatory approach: contradiction and co-option in Burkina Faso', *World Development* 26(12): 2105–2118.

Mills, Mary Beth (1997) 'Contesting the margins of modernity: women, migration, and consumption in Thailand', *American Ethnologist* 24(1): 37–61.

Mills, Mary Beth (1999) *Thai women in the global labor force: consumed desires, contested selves*, New Brunswick, NJ: Rutgers University Press.

Mittelman, James H. (2001) 'Mapping globalisation', *Singapore Journal of Tropical Geography* 22(3): 212–218.

Mohan, Giles (2001) 'Beyond participation: strategies for deeper empowerment', in: Bill Cooke and Uma Kothari (eds) *Participation: the new tyranny?*, London: Zed Books, pp. 154–167.

Mohan, Giles and Stokke, Kristian (2000) 'Participatory development and empowerment: the dangers of localism', *Third World Quarterly* 21(2): 247–268.

Momsen, Janet (1993) 'Women, work and the life course in the rural Caribbean', in: Janice Monk and Cindi Katz (eds) *Full circles: geographies of women over the life course*, London: Routledge, pp. 122–137.

Monk, Janice and Katz, Cindi (1993) 'When in the world are women?', in: Cindi Katz and Janice Monk (eds) *Full circles: geographies of women over the life course*, London: Routledge, pp. 1–26.

Mood, Michelle S. (2005) 'Opportunists, predators and rogues: the role of local state relations in shaping Chinese rural development', *Journal of Agrarian Change* 5(2): 217–250.

Moseley, William G. (2005) 'Global cotton and local environmental management: the political ecology of rich and poor small-holder farmers in southern Mali', *The Geographical Journal* 171(1): 36–55.

Mukherjee, Neela (2004) 'Migrant women from West Bengal – livelihoods, vulnerability, ill-being and well being: some perspectives from the field'. Downloaded from http://www.eldis.org/fulltext/migrantwomen.doc.

Murray, Colin (2002) 'Livelihoods research: transcending boundaries of time and space', *Journal of Southern African Studies* 28(3): 489–509.

Nabi, Rashed un, Datta, Dipankara, Chakrabarty, Subrata, Begum, Masuma and Chaudhury, Nasima Jahan (1999) *Consultation with the poor: participatory poverty assessment in Bangladesh*, NGO Working Group on the World Bank, Bangladesh, June. Downloadable from http://www1.worldbank.org/prem/poverty/voices/reports/national/banglade.pdf.

Nabi, Rashed un, Datta, Dipankar, and Chakrabarty, Subrata (2000) 'Bangladesh: waves of disaster', in: Deepa Narayan and Patti Petesch (eds) *Voices of the poor: from many lands*, New York: Oxford University Press, pp. 113–145. Downloadable from http://www1.worldbank.org/prem/poverty/voices/reports.htm.

Narayan, Deepa and Pritchett, Lant (1999) 'Cents and sociability: household income and social capital in rural Tanzania', *Economic Development and Cultural Change* 47(4): 871–897.

Narayan, Deepa and Petesch, Patti (2000) 'Introduction', in: Deepa Narayan and Patti Petesch (eds) *Voices of the poor: from many lands*, New York: Oxford University Press, pp. 1–15. Downloadable from http://www1.worldbank.org/prem/poverty/voices/reports.htm.

Narayan, Deepa, Chambers, Robert, Shah, Meera K. and Petesch, Patti (2000) *Voices of the Poor: crying out for change*, New York: Oxford University Press. Downloadable from: http://www1.worldbank.org/prem/poverty/voices/reports.htm.

Ninno, Carlo de, Dorosh, Paul A., and Nurul, Islam (2002) 'Reducing vulnerability to natural disasters: lessons from the 1998 floods in Bangladesh', *IDS Bulletin* 33(4): 98–107.

NSOM (2001) *Mongolia: participatory living standards assessment 2000*, summary report prepared for Donor Consultative Group Meeting, Paris, 15–16 May, National Statistical Office of Mongolia and the World Bank. Downloadable from http://www.livelihoods.org/info/docs/plsamong.pdf.

Offe, Claus (1985) 'New social movements: challenging the boundaries of institutional politics', *Social Research* 52(4): 817–868.

Olds, Kris (2001) 'Practices for "process geographies": a view from within and outside the periphery', *Environment and Planning D: Society and Space* 19(2): 127–136.

Ollenu, M.A. (1962) *Principles of customary land law in Ghana*, London: Sweet and Maxwell.

Ong, Aihwa (1987) *Spirits of resistance and capitalist discipline: factory women in Malaysia*, Albany, NY: State University of New York Press.

Opel, Aftab (2005) *Bound for the city: a study of rural to urban labour migration in Afghanistan*, Working Paper, Afghanistan Research and Evaluation Unit, Kabul, Afghanistan (April). Downloaded from http://www.areu.org.af/index.php?option=com_contentandtask=viewandid=41andItemid=86.

O'Rourke, Dara (2004) *Community-driven regulation: balancing development and the environment in Vietnam*, Cambridge, MA: MIT Press.

Osella, Filippo and Osella, Caroline (1999) 'From transience to immanence: consumption, life-cycle and social mobility in Kerala, South India', *Modern Asian Studies* 33(4): 989–1020.

Oslender, Ulrich (2004) 'Fleshing out the geographies of social movements: Colombia's Pacific coast black communities and the "aquatic space"', *Political Geography* 23: 957–985.

Oster, Emily (2005) *Hepatitis B and the case of the missing women*, Harvard University Working Paper, August. Downloaded from http://www.people.fas.harvard.edu/~eoster/hepb.pdf.

Oxfam (2004) *Trading away our rights: women working in global supply chains*, Oxford: Oxfam International. Downloadable from http://www.oxfam.org.uk/what_we_do/issues/trade/trading_rights.htm.

Oxfam (2005) 'The tsunami's impact on women', Oxfam Briefing Notes, March. Downloadable from http://www.oxfam.org.uk/waht_we_do/issues/conflict_disasters/downloads/bn_tsunami_women .pdf.

Painter, Martin (2003) 'The politics of economic restructuring in Vietnam: the case of state-owned enterprise "reform"', *Contemporary Southeast Asia* 25(1): 20–43.

Painter, Martin (2005) 'The politics of state sector reforms in Vietnam: contested agendas and uncertain trajectories', *Journal of Development Studies* 41(2): 261–283.

Peluso, Nancy Lee (1995) 'Whose woods are these? Counter-mapping forest territories in Kalimantan, Indonesia', *Antipode* 27(4): 383–406.

Pieke, Frank N. (2004) 'Contours of an anthropology of the Chinese state: political structure, agency and economic development in rural China', *Journal of the Royal Anthropological Institute NS* 10: 517–538.

Pigg, Stacy Leigh (1992) 'Inventing social categories through place: social representations and development in Nepal', *Comparative Studies in Society and History* 34(3): 491–513.

Pigg, Stacy Leigh (1995) 'The social symbolism of healing in Nepal', *Ethnology* 34(1): 17–36.

Pigg, Stacy Leigh (1996) 'The credible and the credulous: the question of "villagers' beliefs" in Nepal', *Cultural Anthropology* 11(2): 160–201.

Pincus, J. (1996) *Class power and agrarian change: land and labour in rural West Java*, London: Macmillan Press.

Polanyi, K. (1957) *The great transformation: the political and economic origins of our time*, Boston, MA: Beacon Press.

Popkin, Samuel L. (1979) *The rational peasant: the political economy of rural society in Vietnam*, Berkeley, CA: University of California Press.

Porter, Gina and Lyon, Fergus (2006) 'Groups as a means or an end? Social capital and the promotion of cooperation in Ghana', *Environment and Planning D: Society and Space* 24: 249–262.

Potter, Rob (2002) 'Geography and development: "core and periphery"? A reply', *Area* 34(2): 213–214.

Potts, Deborah (1995) 'Shall we go home? Increasing urban poverty in African cities and migration processes', *The Geographical Journal* 161(3): 245–264.

Potts, Deborah (2000) 'Urban unemployment and migrants in Africa: evidence from Harare 1985–1994', *Development and Change* 31(4): 879–910.

Potts, Deborah (2005) 'Counter-urbanisation on the Zambian Copperbelt? Interpretations and implications', *Urban Studies* 42(4): 583–609.

Putnam, Robert, with Leonardi, Robert and Nanetti, Raffaella Y. (1993) *Making democracy work: civic traditions in modern Italy*, Princeton, NJ: Princeton University Press.

Putzel, James (1997) 'Accounting for the dark side of social capital: reading Robert Putnam on democracy', *Journal of International Development* 9(7): 939–949.

Radcliffe, Sarah A. (ed.) (2006) *Culture and development in a globalizing world: geographies, actors and paradigms*, Abingdon, Oxon: Routledge.

Radcliffe, Sarah A. and Laurie, Nina (2006) 'Culture and development: taking culture seriously in development for Andean indigenous people', *Environment and Planning D: Society and Space* 24: 231–248.

Rapley, John (2004) 'Development studies and the post-development critique', *Progress in Development Studies* 4(4): 350–54.

Ravallion, Martin (2001) 'Growth, inequality and poverty: looking beyond averages', *World Development* 29(11): 1803–1815.

Rawski, Thomas G. and Mead, Robert W. (1998) 'On the trail of China's phantom farmers', *World Development* 26(5): 767–781.

Raynolds, Laura T., Myhre, David, McMichael, Philip, Carro-Figueroa, Viviana, and Buttel, Frederick H. (1993) 'The "new" internationalization of agriculture: a reformulation', *World Development* 21(7): 1101–1121.

Reardon, Thomas, Berdegué, Julio A., Timmer, C. Peter, Cabot, Thomas, Mainville, Denise, Flores, Luis, Hernandez, Ricardo, Neven, David, and Balsevich, Fernando (2005) 'Links among supermarkets, wholesalers and small farmers in developing countries: conceptualization and emerging evidence', paper presented at the workshop on *The Future of Small Farms*, Wye, UK, 26–29 June 2005. Downloadable from http://www.ifpri.org/events/seminars/2005/smallfarms/sfproc.asp.

Resurreccion, Bernadette P. (2006) 'Gender, identity and agency in Philippine upland development', *Development and Change* 37(2): 375–400.

Resurreccion, Bernadette, Real, Mary Jane and Pantana, Panadda (2004) 'Officialising strategies and gender in Thailand's water resources sector', *Development in Practice* 14(4): 521–533.

Rigg, Jonathan (2002) 'Rural areas, rural people and the Asian crisis: ordinary people in a globalising world', in: Pietro Paolo Masina (ed.) *Rethinking development in East Asia: from illusory miracle to economic crisis*, Richmond, Surrey: Curzon, pp. 241–260.

Rigg, Jonathan (2003) *Southeast Asia: the human landscape of modernization and development*, London: Routledge.

Rigg, Jonathan (2005) *Living with transition in Laos: market integration in Southeast Asia*, London: Routledge.

Rigg, Jonathan (2006) 'Land, farming, livelihoods, and poverty: rethinking the links in the Rural South', *World Development* 34(1): 180–202.

Rigg, Jonathan (2007) 'Moving lives: migration and livelihoods in the Lao PDR', *Population, Space and Place* 13(3).

Rigg, Jonathan, Grundy-Warr, Carl, Law, Lisa and Tan-Mullins, May (2006) 'The Indian Ocean tsunami: socio-economic impacts in Thailand', *The Geographical Journal* 171(4): 374–379.

Robins, Nick (2002) 'Loot: in search of the East India Company, the world's first transnational corporation', *Environment and Urbanization* 14(1): 79–83.

Robinson, Jenny (2002) 'Global and world cities: a view from off the map', *International Journal of Urban and Regional Research* 26(3): 531–554.

Robinson, Jenny (2003) 'Postcolonialising geography: tactics and pitfalls', *Singapore Journal of Tropical Geography* 24(3): 273–289.

Røe, P.G. (2000) 'Qualitative research on intra-urban travel: an alternative approach', *Journal of Transport Geography* 8(2): 99–106.

Rogaly, Ben and Coppard, Daniel (2003) '"They used to go to eat, now they go to earn": the changing meanings of seasonal migration from Puruliya District in West Bengal, India', *Journal of Agrarian Change* 3(3): 395–433.

Rogaly, Ben and Rafique, Abdur (2003) 'Struggling to save cash: seasonal migration and vulnerability in West Bengal, India', *Development and Change* 34(4): 659–681.

Routledge, Paul (2003) 'Convergence space: process geographies of grassroots globalization networks', *Transactions of the Institute of British Geographers* New Series, 28: 333–349.

Russell, Margo (1993) 'Are households univeral? On misunderstanding domestic groups in Swaziland', *Development and Change* 24(4): 755–785.

Sabates-Wheeler, Rachel, Sabates, Ricardo, and Castaldo, Adriana (2005) *Tackling poverty-migration linkages: evidence from Ghana and Egypt*, Working Paper T14, Development Research Centre on Migration, Globalisation and Poverty, Sussex: Institute of Development Studies. Downloaded from http://www.migrationdrc.org/publications/working_papers/WP-T14.pdf.

Sadler, David (2004) 'Anti-corporate campaigning and corporate "social" responsibility: towards alternative spaces of citizenship?', *Antipode* 36(5): 851–870.

Sanderson, Steven E. (1986) *The transformation of Mexican agriculture: international structure and the politics of rural change*, Princeton, NJ: Princeton University Press.

Saptari, Ratna and Elmhirst, Rebecca (2004) 'Studying labour in Southeast Asia: reflections on structures and processes', in: Rebecca Elmhirst and Ratna Saptari (eds) *Labour in Southeast Asia: local processes in a globalised world*, London: Routledge, pp. 15–46.

Satterthwaite, David (2003) 'The Millennium Development Goals and urban poverty reduction: great expectations and nonsense statistics', *Environment and Urbanization* 15(2): 181–190.

Schirato, Tony and Webb, Jen (2003) *Understanding globalization*, London: Sage.

Schmitz, Hubert (1995) 'Small shoemakers and Fordist giants: tale of a supercluster', *World Development* 23(1): 9–28.

Schönwälder, Gerd (1997) 'New democratic spaces at the grassroots? Popular participation in Latin American local governments', *Development and Change* 28: 753–770.

Schuurman, Frans J. (2003) 'Social capital: the politico-emancipatory potential of a disputed concept', *Third World Quarterly* 24(6): 991–1010.

Scoones, Ian (1998) *Sustainable rural livelihoods: a framework for analysis*, IDS Working Paper 72, Brighton, Sussex: Institute of Development Studies. Downloadable from http://www.ids.ac.uk/ids/bookshop/wp.html.

Scoones, Ian and Wolmer, William (2003) 'Livelihoods in crisis: challenges for rural development in Southern Africa', *IDS Bulletin* 34(3): 1–14.

Scott, James C. (1976) *The moral economy of the peasant: rebellion and subsistence in Southeast Asia*, New Haven, CT: Yale University Press.

Scott, James C. (1985) *Weapons of the weak: everyday forms of peasant resistance*, New Haven, CT: Yale University Press.

Scott, James C. (1998) *Seeing like a state: how certain schemes to improve the human condition have failed*, New Haven, CT: Yale University Press.

Sen, Amartya (1981) *Poverty and famines: an essay on entitlement and deprivation*, Oxford: Oxford University Press.

Sen, Amartya (1990a) 'More than 100 million women are missing', *New York Review of Books* 37(20). Downloaded from http://ucatlas.ucsc.edu/gender/Sen100M.html#fn4.

Sen, Amartya (1990b) 'Gender and cooperative conflicts', in: Irene Tinker (ed.) *Persistent inequalities: women and world development*, New York: Oxford University Press, pp. 123–149.

Sen, Amartya (2003) 'Editorial: Missing women – revisited', *British Medical Journal* 327: 1297–1298. Downloaded from http://bmj.bmjjournals.com/cgi/content/full/327/7427/1297.

Sen, Samita (1999) *Women and labour in late colonial India: the Bengal jute industry*, Cambridge: Cambridge University Press.

Sheller, Mimi and Urry, John (2006) 'The new mobility paradigm', *Environment and Planning* A 38(2): 207–226.

Shen, Xiaoping and Ma, Lawrence J.C. (2005) 'Privatization of rural industry and de facto urbanization from below in southern Jiangsu, China', *Geoforum* 36: 761–777.

Sherman, D. George (1990) *Rice, rupees and ritual: economy and society among the Samosir Batak of Sumatra*, Stanford, CA: Stanford University Press.

Silvey, Rachel (2001) 'Migration under crisis: household safety nets in Indonesia's economic collapse', *Geoforum* 32: 33–45.

Silvey, Rachel (2003) 'Spaces of protest: gendered migration, social networks, and labor activism in West Java, Indonesia', *Political Geography* 22: 129–155.

Silvey, Rachel (2004) 'Transnational domestication: state power and Indonesian migrant women in Saudi Arabia', *Political Geography* 23: 245–264.

Silvey, Rachel (2006) 'Consuming the transnational family: Indonesian migrant domestic workers to Saudi Arabia', *Global Networks* 6: 23–40.

Silvey, Rachel and Elmhirst, Rebecca (2003) 'Engendering social capital: women workers and rural–urban networks in Indonesia's crisis', *World Development* 31(5): 865–879.

Simard, Paule and De Koninck, Maria (2001) 'Environment, living spaces, and health: compound-organisation practices in a Bamako squatter settlement, Mali', *Gender and Development* 9(2): 28–39.

Singh, Sukhpal (2002) 'Contracting out solutions: political economy of contract farming in the Indian Punjap', *World Development* 30(9): 1621–1638.

Slater, David (1997) 'Geopolitical imaginations across the North–South divide: issues of difference, development and power', *Political Geography* 16(8): 631–653.

Slater, Rachel (2002) 'Differentiation and diversification: changing livelihoods in Qwaqwa, South Africa, 1970–2000', *Journal of Southern African Studies* 28(3): 599–614.

SLSA (2003) 'Livelihood dynamics: rural Mozambique, South Africa and Zimbabwe', *IDS Bulletin* 34(3): 15–30.

Stockdale, Aileen (2004) 'Rural out-migration: community consequences of individual migrant experiences', *Sociologia Ruralis* 44(2): 167–194.

Stockdale, Aileen (2006) 'Migration: pre-requisite for rural economic regeneration?', *Journal of Rural Studies* 22(3): 354–366.

Stoddart, D.R. (1996) 'Correspondence, by Vince Gardiner; D.R. Stoddart; K.J. Gregory; R.J. Bennett; Keith Richards; Neil Wrigley', *The Geographical Journal* 162(3): 354–357.

Sunshine, Russell B. (1995) *Managing foreign investment: lessons from Laos*, Honolulu: East–West Center.

Tanabe, Shigeharu and Keyes, Charles F. (2002) 'Introduction', in: Shigeharu Tanabe and Charles F. Keyes (eds) *Cultural crisis and social memory: modernity and identity in Thailand and Laos*, London: RoutledgeCurzon, pp. 1–39.

Taylor, P.J. (1993) 'Full circle or new meaning for the global?', in: R.J. Johnston (ed.) *The challenge for geography: a changing world, a changing discipline*, Oxford: Blackwells, pp. 181–197.

Taylor, P.J. (1999) 'Places, spaces, and Macy's: place-space tensions in the political geographies of modernities', *Progress in Human Geography* 23(1): 7–26.

Taylor, P.J., Watts, M.J. and Johnston, R.J. (2001) 'Geography/Globalization', *Research Bulletin* 40, http://www.lboro.ac.uk/gawc/rb/rb40.html.

Thangarajah, C.Y. (2004) 'Veiled constructions: conflict, migration and modernity in eastern Sri Lanka', in: Katy Gardner and Filippo Osella (eds) *Migration, modernity and social transformation in South Asia*, Contributions to Indian Sociology, Occasional Studies 11, New Delhi: Sage, pp. 141–162.

Thayer, Carl (1992) 'Political reform in Vietnam: doi moi and the emergence of civil society', in: Robert F. Miller (ed.) *The development of civil society in communist systems*, Sydney: Allen and Unwin, pp. 110–129.

Thin, Neil with Good, Tony and Hodgson, Rebecca (1997) *Social development policies, results and learning: a multi-agency review*, DfID: Social Development Department, SD SCOPE (Social Development Systems for Coordinated Poverty Eradication). Downloaded from http://www.dfid.gov.uk/pubs/files/sddmulti.pdf.

Thomas, David S.G. and Twyman, Chasca (2005) 'Equity and justice in climate change adaptation amongst natural-resource-dependent societies', *Global Environmental Change* 15: 115–124.

Thomas, Nicholas (1994) *Colonialism's culture: anthropology, travel and government*, Cambridge: Polity Press.

Thompson, Eric C. (2002) 'Migrant subjectivities and narratives of the *kampung* in Malaysia', *Sojourn* 17(1): 52–75.

Thompson, Eric C. (2003) 'Malay male migrants: negotiating contested identities in Malaysia', *American Ethnologist* 30(3): 418–438.

Thompson, Eric C. (2004) 'Rural villages and socially urban spaces in Malaysia', *Urban Studies* 41(12): 2357–2376.

Thorbecke, Erik and Nissanke, Machiko (2006) 'Introduction: the impact of globalization on the world's poor', *World Development* 34(8): 1333–1337.

Thrift, Nigel and Walling, Des (2000) 'Geography in the United Kingdom 1996–2000', *The Geographical Journal* 166(2): 96–124.

Townsend, Janet G. (1999) 'Are non-governmental organisations a transnational community?', *Journal of International Development* 11: 613–623.

Townsend, Janet G., Porter, Gina and Mawdsley, Emma (2004) 'Creating spaces of resistance: development NGOs and their clients in Ghana, India and Mexico', *Antipode* 36(5): 871–889.

Turner, Sarah and Phuong An Nguyen (2005) 'Young entrepreneurs, social capital and *doi moi* in Hanoi, Vietnam', *Urban Studies* 42(10): 1693–1710.

Twyman, Chasca, Sporton, Deborah and Thomas, David S.G. (2004) '"Where is the life in farming?": the viability of smallholder farming on the margins of the Kalahari, Southern Africa', *Geoforum* 35: 69–85.

Vandergeest, Peter (1996) 'Mapping nature: territorialization of forest rights in Thailand', *Society and Natural Resources* 9: 159–175.

Vandergeest, Peter (2003) 'Land to some tillers: development-induced displacement in Laos', *International Social Science Journal* 175(March): 47–56.

Vandergeest, Peter and Peluso, Nancy Lee (1995) 'Territorialization and state power in Thailand', *Theory and Society* 24(3): 385–426.

Waddington, Hugh and Sabates-Wheeler, Rachel (2003) *How does poverty affect migration choice? A review of the literature*, Working Paper T3, Development Research Centre on Migration, Globalisation and Poverty, Sussex: Institute of Development Studies. Downloaded from http://www.migrationdrc.org/publications/working_papers/WP-T3.pdf.

Wadley, Reed L. (2003) 'Lines in the forest: internal territorialization and local accommodation in West Kalimantan, Indonesia (1865–1979)', *South East Asia Research* 11(1): 91–112.

Wallerstein, Immanuel (1974) *The modern world-system, volume 1: capitalist agriculture and the origins of the European world-economy in the sixteenth century*, New York: Academic Press.

Wallerstein, Immanuel (1980) *The modern world-system, volume 2: Mercantilism and the consolidation of the European world-economy, 1600–1750*, New York: Academic Press.

Wallerstein, Immanuel (1989) *The modern world-system, volume 3: The second era of great expansion of the capitalist world-economy, 1730–1840s*, New York: Academic Press.

Ward, Neil and Almås, Reidar (1997) 'Explaining change in the international agro-food system', *Review of International Political Economy* 4(4): 611–629.

Warouw, Nicolaas (2003) 'Keeping up appearances: manufacturing workers in Tangerang make a special effort to look good', *Inside Indonesia* 75 (July–September). Downloaded from http://www.insideindonesia.org/edit75/p25warouw.html.

Watts, Michael J. (1992) 'Living under contract: work, production politics, and the manufacture of discontent in a peasant society', in: Allan Pred and Michael J. Watts (eds) *Reworking modernity: capitalisms and symbolic discontent*, New Brunswick, NJ: Rutgers University Press, pp. 65–105.

Watts, Michael J. (1994) 'Life under contract: contract farming, agrarian restructuring, and flexible accumulation', in: Peter D. Little and Michael J. Watts (eds) *Living under contract: contract farming and agrarian transformation in sub-Saharan Africa*, Madison, WI: University of Wisconsin Press, pp. 21–77.

Watts, Michael J. and Goodman, David (1997) 'Agrarian questions: global appetite, local metabolism – nature, culture, and industry in *fin-de-siècle* agro-food systems', in: David Goodman and Michael J. Watts (eds) *Globalising food: agrarian questions and global restructuring*, London: Routledge, pp. 1–32.

Webber, Michael and Wang, Mark Y.L. (2004) 'Markets in the Chinese countryside: the case of "rich Wang's village"', *Geoforum* 36: 720–734.

White, B. and Wiradi, G. (1989) Agrarian and nonagrarian bases of inequality in nine Javanese villages. In G. Hart, A. Turton and B. White (eds) *Agrarian transformations: local processes and the state in Southeast Asia*, Berkeley, CA: University of California Press, pp. 266–302.

White, S.A., Nair, K.S., and Ascroft, J. (1994) 'Introduction: the concept of participation: transforming rhetoric to reality', in: S.A. White, K.S. Nair and J. Ascroft (eds) *Participatory communication: working for change and development*, New Delhi: Sage, pp. 15–32.

White, Sarah C. (1996) 'Depoliticising development: the uses and abuses of participation', *Development in Practice* 6(1): 6–15.

Whitehead, Ann (2002) 'Tracking livelihood change: theoretical, methodological and empirical perspectives from North-East Ghana', *Journal of Southern African Studies* 28(3): 575–598.

Williams, Glyn (1997) 'State, discourse, and development in India: the case of West Bengal's *Panchayati Raj*', *Environment and Planning A* 29(12): 2099–2112.

Wilson, Godfrey (1941) *An essay on the economics of detribalization in northern Rhodesia*, Rhodes-Livingstone Paper no. 6.

Wolf, Diane Lauren (1992) *Factory daughters: gender, household dynamics, and rural industrialization in Java*, Berkeley, CA: University of California Press.

Wolf, Diane Lauren (2000) 'Beyond women and the household in Java: re-examining the boundaries', in: Juliette Koning, Marleen Nolten, Janet Rodenburg and Ratna Saptari (eds) *Women and households in Indonesia: cultural notions and social practices*, Richmond, Surrey: Curzon, pp.85–100.

Wolf, Eric R. (1967) 'Closed corporate peasant communities in Mesoamerica and Central Java', in:

Jack M. Potter, May N. Diaz and George M. Foster (eds) *Peasant society: a reader*, Boston, MA: Little, Brown, pp. 230–246. (First published in 1957 in *South Western Journal of Anthropology* 13(1): 1–18)

Wolf, Eric R. (1982) *Europe and the people without history*, Berkeley, CA: University of California Press.

Wolf, Martin (2004) *Why globalization works: the case for the global market economy*, New Haven, CT: Yale University Press.

World Bank (1999) *Methodology guide: consultations with the poor*, Poverty Group, PREM, Washington, DC. Downloadable from http://www1.worldbank.org/prem/poverty/voices/reports/method/method. pdf.

World Bank (2001) *World development report 2000/2001: attacking poverty*, New York: Oxford University Press.

World Commission on Environment and Development (1987) *Our common future*, Oxford: Oxford University Press.

Wright, Melissa W. (2003) 'Factory daughters and Chinese modernity: a case from Dongguan', *Geoforum* 34: 291–301.

Yan Hairong (2003) 'Spectralization of the rural: reinterpreting the labor mobility of rural young women in post-Mao China', *American Ethnologist* 30(4): 578–596.

Yeung, Henry W-C. (1998) 'Capital, state and space: contesting the borderless world', *Transactions of the Institute of British Geographers* New Series, 23(3): 291–309.

Yeung, Henry W-C. (2004) *Chinese capitalism in a global era: towards hybrid capitalism*, London: Routledge.

Yeung, Henry W-C. and Lim, George C.S. (2003) 'Theorizing economic geographies of Asia', *Economic Geography* 79(2): 107–128.

Zamora, M. (2004) 'La rápida expansion de los Supermercados en Ecuador y sus efectos sobre las cadenas de lácteos y papas', report for the Regoverning Markets Project, September.

Zoomers, Annelies (1999) 'Livelihood strategies and development interventions in the Southern Andes of Bolivia: contrasting views on development', *Cuadernos del Cedla no. 3*. Downloadable from http://www.cedla.uva.nl/10_about/PDF_files_about/boliviacderno.pdf.

Index